AFOQT Math
in 10 Days!

The Most Effective
AFOQT Math Crash
Course

By

Reza Nazari

About Effortless Math Education

Effortless Math Education operates the www.effortlessmath.com website, which prepares and publishes Test prep and Mathematics learning resources. Effortless Math authors' team strives to prepare and publish the best quality Mathematics learning resources to make learning Math easier for all. We Help Students Learn to Love Mathematics.

ISBN: 978-1-64612-261-5

Published by: **Effortless Math Education**

for Online Math Practice Visit www.EffortlessMath.com

AFOQT MATH in 10 Days

The ultimate crash course to help you ace the AFOQT Math test!

The perfect quick-review study guide for students of every level, *AFOQT Math in 10 Days* is the top choice for AFOQT Math test takers who want to make the most of their study time and earn a high score. Designed for the student who's running out of time, this book is the perfect last-minute solution that covers only the math concepts and topics tested on the exam, so you can save your valuable study time.

Written by a top AFOQT Math instructor and test prep expert, this quick study guide gives you the critical math concepts that will matter most on exam day. It relies on the author's extensive analysis of the test's structure and content. By following his advice, you can hone your math skills, overcome exam anxiety, and boost your score.

Here's why more teachers and test takers turn to this AFOQT Math crash course:

- Content 100% aligned with the **2020 AFOQT test**

- Written by a top AFOQT Math instructor and test expert

- **Targeted Review** - study only what you need to know

- **Step-by-step guide** for all AFOQT Math topics

- Abundant Math skills building exercises to help test-takers approach unfamiliar question types

- **2 full-length practice tests** (featuring new question types) with detailed answers

- And much more!

When it's crucial crunch time and your AFOQT Math test is just around the corner, you only need *AFOQT Math in 10 Days*! Practice Your Way to Excellence.

Ideal for self-study and classroom usage!

About the Author

Reza Nazari is the author of more than 100 Math learning books including:

- ❖ **Math and Critical Thinking Challenges:** For the Middle and High School Student
- ❖ ACT Math in 30 Days
- ❖ ASVAB Math Workbook 2020
- ❖ Effortless Math Education Workbooks
- ❖ and many more Mathematics books

Reza is also an experienced Math instructor and a test–prep expert who has been tutoring students since 2008. Reza is the founder of Effortless Math Education, a tutoring company that has helped many students raise their standardized test scores––and attend the colleges of their dreams. Reza provides an individualized custom learning plan and the personalized attention that makes a difference in how students view math.

You can contact Reza via email at:
reza@EffortlessMath.com

Contents

DAY 1

Fractions and Mixed Numbers   1

DAY 2

Decimals and Integers   15

DAY 3

Ratios, Proportions and Percent   29

Contents

Contents

Fractions and Mixed Numbers

Math topics that you'll learn in this chapter:

1. Simplifying Fractions

2. Adding and Subtracting Fractions

3. Multiplying and Dividing Fractions

4. Adding Mixed Numbers

5. Subtracting Mixed Numbers

6. Multiplying Mixed Numbers

7. Dividing Mixed Numbers

1

Simplifying Fractions

☆ A fraction contains two numbers separated by a bar between them. The bottom number, called the denominator, is the total number of equally divided portions in one whole. The top number, called the numerator, is how many portions you have. And the bar represents the operation of division.

☆ Simplifying a fraction means reducing it to the lowest terms. To simplify a fraction, evenly divide both the top and bottom of the fraction by 2, 3, 5, 7, etc.

☆ Continue until you can't go any further.

Examples:

Example 1. *Simplify* $\frac{16}{24}$

Solution: To simplify $\frac{16}{24}$, find a number that both 16 and 24 are divisible by. Both are divisible by 8. Then: $\frac{16}{24} = \frac{16 \div 8}{24 \div 8} = \frac{2}{3}$

Example 2. *Simplify* $\frac{36}{96}$

Solution: To simplify $\frac{36}{96}$, find a number that both 36 and 96 are divisible by. Both are divisible by 6 and 12. Then: $\frac{36}{96} = \frac{36 \div 6}{96 \div 6} = \frac{6}{16}$, 6 and 16 are divisible by 2, then: $\frac{6}{16} = \frac{3}{8}$ or $\frac{36}{96} = \frac{36 \div 12}{96 \div 12} = \frac{3}{8}$

Example 3. *Simplify* $\frac{43}{129}$

Solution: To simplify $\frac{43}{129}$, find a number that both 43 and 129 are divisible by. Both are divisible by 43, then: $\frac{43}{129} = \frac{43 \div 43}{129 \div 43} = \frac{1}{3}$

Adding and Subtracting Fractions

✬ For "like" fractions (fractions with the same denominator), add or subtract the numerators (top numbers) and write the answer over the common denominator (bottom numbers).

✬ Adding and Subtracting fractions with the same denominator:

$\frac{a}{b} + \frac{c}{b} = \frac{a+c}{b}, \frac{a}{b} - \frac{c}{b} = \frac{a-c}{b}$

✬ Find equivalent fractions with the same denominator before you can add or subtract fractions with different denominators.

✬ Adding and Subtracting fractions with different denominators:

$\frac{a}{b} + \frac{c}{d} = \frac{ad+bc}{bd}, \frac{a}{b} - \frac{c}{d} = \frac{ad-bc}{bd}$

Examples:

Example 1. Find the sum. $\frac{3}{4} + \frac{2}{3} =$

Solution: These two fractions are "unlike" fractions. (they have different denominators). Use this formula: $\frac{a}{b} + \frac{c}{d} = \frac{ad+cb}{bd}$

Then: $\frac{3}{4} + \frac{2}{3} = \frac{(3)(3)+(4)(2)}{4 \times 3} = \frac{9+8}{12} = \frac{17}{12}$

Example 2. Find the difference. $\frac{4}{7} - \frac{2}{5} =$

Solution: For "unlike" fractions, find equivalent fractions with the same denominator before you can add or subtract fractions with different denominators. Use this formula: $\frac{a}{b} - \frac{c}{d} = \frac{ad-bc}{bd}$

$\frac{4}{7} - \frac{2}{5} = \frac{(4)(5)-(2)(7)}{7 \times 5} = \frac{20-14}{35} = \frac{6}{35}$

Multiplying and Dividing Fractions

☆ **Multiplying fractions:** multiply the top numbers and multiply the bottom numbers. Simplify if necessary. $\frac{a}{b} \times \frac{c}{d} = \frac{a \times c}{b \times d}$

☆ **Dividing fractions:** Keep, Change, Flip

☆ Keep the first fraction, change the division sign to multiplication, and flip the numerator and denominator of the second fraction. Then, solve!

$$\frac{a}{b} \div \frac{c}{d} = \frac{a}{b} \times \frac{d}{c} = \frac{a \times d}{b \times c}$$

Examples:

Example 1. Multiply. $\frac{3}{4} \times \frac{2}{5} =$

Solution: Multiply the top numbers and multiply the bottom numbers.
$\frac{3}{4} \times \frac{2}{5} = \frac{3 \times 2}{4 \times 5} = \frac{6}{20}$, now, simplify: $\frac{6}{20} = \frac{6 \div 2}{20 \div 2} = \frac{3}{10}$

Example 2. Solve. $\frac{2}{3} \div \frac{3}{7} =$

Solution: Keep the first fraction, change the division sign to multiplication, and flip the numerator and denominator of the second fraction.
Then: $\frac{2}{3} \div \frac{3}{7} = \frac{2}{3} \times \frac{7}{3} = \frac{2 \times 7}{3 \times 3} = \frac{14}{9}$

Example 3. Calculate. $\frac{6}{5} \times \frac{2}{3} =$

Solution: Multiply the top numbers and multiply the bottom numbers.
$\frac{6}{5} \times \frac{2}{3} = \frac{6 \times 2}{5 \times 3} = \frac{12}{15}$, simplify: $\frac{12}{15} = \frac{12 \div 3}{15 \div 3} = \frac{4}{5}$

Example 4. Solve. $\frac{4}{5} \div \frac{3}{8} =$

Solution: Keep the first fraction, change the division sign to multiplication, and flip the numerator and denominator of the second fraction.
Then: $\frac{4}{5} \div \frac{3}{8} = \frac{4}{5} \times \frac{8}{3} = \frac{4 \times 8}{5 \times 3} = \frac{32}{15}$

Adding Mixed Numbers

Use the following steps for adding mixed numbers:

☆ Add whole numbers of the mixed numbers.

☆ Add the fractions of the mixed numbers.

☆ Find the Least Common Denominator (LCD) if necessary.

☆ Add whole numbers and fractions.

☆ Write your answer in lowest terms.

Examples:

Example 1. Add mixed numbers. $3\frac{2}{3} + 1\frac{2}{5} =$

Solution: Let's rewriting our equation with parts separated, $3\frac{2}{3} + 1\frac{2}{5} = 3 + \frac{2}{3} + 1 + \frac{2}{5}$. Now, add whole number parts: $3 + 1 = 4$. Add the fraction parts $\frac{2}{3} + \frac{2}{5}$. Rewrite to solve with the equivalent fractions. $\frac{2}{3} + \frac{2}{5} = \frac{10}{15} + \frac{6}{15} = \frac{16}{15}$. The answer is an improper fraction (numerator is bigger than denominator). Convert the improper fraction into a mixed number: $\frac{16}{15} = 1\frac{1}{15}$. Now, combine the whole and fraction parts: $4 + 1\frac{1}{15} = 5\frac{1}{15}$.

Example 2. Find the sum. $2\frac{1}{2} + 1\frac{3}{5} =$

Solution: Rewriting our equation with parts separated, $2 + \frac{1}{2} + 1 + \frac{3}{5}$. Add the whole number parts: $2 + 1 = 3$. Add the fraction parts: $\frac{1}{2} + \frac{3}{5} = \frac{5}{10} + \frac{6}{10} = \frac{11}{10}$. Convert the improper fraction into a mixed number: $\frac{11}{10} = 1\frac{1}{10}$. Now, combine the whole and fraction parts: $3 + 1\frac{1}{10} = 4\frac{1}{10}$.

Subtracting Mixed Numbers

Use these steps for subtracting mixed numbers.

☆ Convert mixed numbers into improper fractions. $a\frac{c}{b} = \frac{ab+c}{b}$

☆ Find equivalent fractions with the same denominator for unlike fractions. (fractions with different denominators)

☆ Subtract the second fraction from the first one. $\frac{a}{b} - \frac{c}{d} = \frac{ad-bc}{bd}$

☆ Write your answer in lowest terms.

☆ If the answer is an improper fraction, convert it into a mixed number.

Examples:

Example 1. Subtract. $2\frac{1}{5} - 1\frac{2}{3} =$

Solution: Convert mixed numbers into fractions:

$2\frac{1}{5} = \frac{2\times5+1}{5} = \frac{11}{5}$ and $1\frac{2}{3} = \frac{1\times3+2}{3} = \frac{5}{3}$

These two fractions are "unlike" fractions. (they have different denominators).

Find equivalent fractions with the same denominator. Use this formula:

$$\frac{a}{b} - \frac{c}{d} = \frac{ad - bc}{bd}$$

$\frac{11}{5} - \frac{5}{3} = \frac{(11)(3)-(5)(5)}{5\times3} = \frac{33-25}{15} = \frac{8}{15}$

Example 2. Find the difference. $2\frac{3}{7} - 1\frac{4}{5} =$

Solution: Convert mixed numbers into fractions:

$2\frac{3}{7} = \frac{2\times7+3}{7} = \frac{17}{7}$ and $1\frac{4}{5} = \frac{1\times5+4}{5} = \frac{9}{5}$

Then: $2\frac{3}{7} - 1\frac{4}{5} = \frac{17}{7} - \frac{9}{5} = \frac{(17)(5)-(9)(7)}{7\times5} = \frac{85-63}{35} = \frac{22}{35}$

Multiplying Mixed Numbers

Use the following steps for multiplying mixed numbers:

☆ Convert the mixed numbers into fractions. $a\frac{c}{b} = a + \frac{c}{b} = \frac{ab+c}{b}$

☆ Multiply fractions. $\frac{a}{b} \times \frac{c}{d} = \frac{a \times c}{b \times d}$

☆ Write your answer in lowest terms.

☆ If the answer is an improper fraction (numerator is bigger than denominator), convert it into a mixed number.

Examples:

Example 1. Multiply. $3\frac{1}{3} \times 2\frac{3}{5} =$

Solution: Convert mixed numbers into fractions,
$3\frac{1}{3} = \frac{3 \times 3 + 1}{3} = \frac{10}{3}$ and $2\frac{3}{5} = \frac{2 \times 5 + 3}{5} = \frac{13}{5}$. Apply the fractions rule for multiplication:
$\frac{10}{3} \times \frac{13}{5} = \frac{10 \times 13}{3 \times 5} = \frac{130}{15} = \frac{130 \div 5}{15 \div 5} = \frac{26}{3}$
The answer is an improper fraction. Convert it into a mixed number. $\frac{26}{3} = 8\frac{2}{3}$

Example 2. Multiply. $2\frac{3}{4} \times 4\frac{3}{7} =$

Solution: Converting mixed numbers into fractions, $2\frac{3}{4} \times 4\frac{3}{7} = \frac{11}{4} \times \frac{31}{7}$
Apply the fractions rule for multiplication: $\frac{11}{4} \times \frac{31}{7} = \frac{11 \times 31}{4 \times 7} = \frac{341}{28} = 12\frac{5}{28}$

Example 3. Find the product. $4\frac{3}{5} \times 3\frac{4}{6} =$

Solution: Convert mixed numbers to fractions: $4\frac{3}{5} = \frac{23}{5}$ and $3\frac{4}{6} = \frac{22}{6} = \frac{22 \div 2}{6 \div 2} = \frac{11}{3}$.

Multiply two fractions:

$\frac{23}{5} \times \frac{11}{3} = \frac{23 \times 11}{5 \times 3} = \frac{253}{15} = 16\frac{13}{15}$

Dividing Mixed Numbers

Use the following steps for dividing mixed numbers:

★ Convert the mixed numbers into fractions. $a\frac{c}{b} = a + \frac{c}{b} = \frac{ab+c}{b}$

★ Divide fractions: Keep, Change, Flip: Keep the first fraction, change the division sign to multiplication, and flip the numerator and denominator of the second fraction. Then, solve! $\frac{a}{b} \div \frac{c}{d} = \frac{a}{b} \times \frac{d}{c} = \frac{a \times d}{b \times c}$

★ Write your answer in lowest terms.

★ If the answer is an improper fraction (numerator is bigger than denominator), convert it into a mixed number.

Examples:

Example 1. Solve. $2\frac{2}{3} \div 1\frac{1}{2} =$

Solution: Convert mixed numbers into fractions:
$2\frac{2}{3} = \frac{2\times3+2}{3} = \frac{8}{3}$ and $1\frac{1}{2} = \frac{1\times2+1}{2} = \frac{3}{2}$
Keep, Change, Flip: $\frac{8}{3} \div \frac{3}{2} = \frac{8}{3} \times \frac{2}{3} = \frac{8\times2}{3\times3} = \frac{16}{9}$. The answer is an improper fraction.
Convert it into a mixed number: $\frac{16}{9} = 1\frac{7}{9}$

Example 2. Solve. $4\frac{2}{3} \div 1\frac{3}{5} =$

Solution: Convert mixed numbers to fractions, then solve:
$4\frac{2}{3} \div 1\frac{3}{5} = \frac{14}{3} \div \frac{8}{5} = \frac{14}{3} \times \frac{5}{8} = \frac{70}{24} = 2\frac{11}{12}$

Example 3. Solve. $3\frac{2}{5} \div 2\frac{1}{3} =$

Solution: Converting mixed numbers to fractions: $3\frac{2}{5} \div 2\frac{1}{3} = \frac{17}{5} \div \frac{7}{3}$
Keep, Change, Flip: $\frac{17}{5} \div \frac{7}{3} = \frac{17}{5} \times \frac{3}{7} = \frac{17\times3}{5\times7} = \frac{51}{35} = 1\frac{16}{35}$

Day 1: Practices

✎ Simplify each fraction.

1) $\frac{2}{8} =$

2) $\frac{5}{15} =$

3) $\frac{12}{36} =$

4) $\frac{65}{120} =$

✎ Find the sum or difference.

5) $\frac{3}{10} + \frac{2}{10} =$

6) $\frac{4}{5} + \frac{1}{10} =$

7) $\frac{3}{4} + \frac{6}{20} =$

8) $\frac{4}{9} - \frac{1}{9} =$

9) $\frac{3}{8} - \frac{1}{6} =$

10) $\frac{9}{21} - \frac{2}{7} =$

✎ Find the products or quotients.

11) $\frac{3}{4} \div \frac{9}{12} =$

12) $\frac{7}{10} \div \frac{21}{20} =$

13) $\frac{12}{21} \div \frac{3}{7} =$

14) $\frac{12}{5} \times \frac{10}{24} =$

15) $\frac{33}{36} \times \frac{3}{4} =$

16) $\frac{7}{9} \times \frac{1}{3} =$

✎ Find the sum.

17) $3\frac{1}{2} + 1\frac{3}{4} =$

18) $4\frac{1}{8} + 2\frac{7}{8} =$

19) $4\frac{1}{2} + 2\frac{3}{8} =$

20) $1\frac{2}{21} + 4\frac{4}{7} =$

21) $6\frac{3}{5} + 1\frac{2}{3} =$

22) $2\frac{3}{11} + 3\frac{1}{2} =$

✎ Find the difference.

23) $6\frac{2}{3} - 4\frac{1}{3} =$

24) $5\frac{2}{5} - 3\frac{1}{5} =$

25) $8\frac{1}{2} - 3\frac{1}{4} =$

26) $7\frac{2}{3} - 2\frac{1}{6} =$

✍ Find the products.

27) $1\frac{2}{3} \times 2\frac{3}{4} =$

28) $1\frac{1}{6} \times 1\frac{3}{5} =$

29) $4\frac{1}{2} \times 1\frac{2}{3} =$

30) $2\frac{1}{2} \times 4\frac{4}{5} =$

31) $2\frac{1}{5} \times 4\frac{1}{2} =$

32) $1\frac{1}{9} \times 2\frac{3}{5} =$

✍ Solve.

33) $3\frac{1}{3} \div 1\frac{2}{3} =$

34) $4\frac{2}{3} \div 2\frac{1}{2} =$

35) $6\frac{1}{5} \div 2\frac{1}{3} =$

36) $2\frac{2}{3} \div 1\frac{4}{9} =$

37) $4\frac{1}{6} \div 2\frac{1}{8} =$

38) $3\frac{2}{5} \div 1\frac{5}{4} =$

39) A pizza cut into 6 parts. David and his sister Sara ordered two pizzas. David ate $\frac{1}{3}$ of his pizza and Eva ate $\frac{1}{2}$ of her pizza. What part of the two pizzas was left?

40) Jake is preparing to run a marathon. He runs $9\frac{1}{3}$ miles on Saturday and two times that many on Monday and Wednesday. Jake wants to run a total of 50 miles this week. How many more miles does he need to run?

41) Last week 21,000 fans attended a football match. This week four times as many bought tickets, but one third of them cancelled their tickets. How many are attending this week?

42) In a bag of small balls $\frac{1}{2}$ are black, $\frac{1}{4}$ are white, $\frac{1}{8}$ are red and the remaining 16 blue. How many balls are white?

Day 1: Answers

1) $\frac{2}{8} = \frac{2 \div 2}{8 \div 2} = \frac{1}{4}$

2) $\frac{5}{15} = \frac{5 \div 5}{15 \div 5} = \frac{1}{3}$

3) $\frac{12}{36} = \frac{12 \div 12}{36 \div 12} = \frac{1}{3}$

4) $\frac{65}{120} = \frac{65 \div 5}{120 \div 5} = \frac{13}{24}$

5) $\frac{3}{10} + \frac{2}{10} = \frac{3+2}{10} = \frac{5}{10} = \frac{5 \div 5}{10 \div 5} = \frac{1}{2}$

6) $\frac{4}{5} + \frac{1}{10} = \frac{4 \times 2}{5 \times 2} + \frac{1}{10} = \frac{8}{10} + \frac{1}{10} = \frac{8+1}{10} = \frac{9}{10}$

7) $\frac{3}{4} + \frac{6}{20} = \frac{3 \times 5}{4 \times 5} + \frac{6}{20} = \frac{15}{20} + \frac{6}{20} = \frac{21}{20}$

8) $\frac{4}{9} - \frac{1}{9} = \frac{4-1}{9} = \frac{3}{9} = \frac{3 \div 3}{9 \div 3} = \frac{1}{3}$

9) $\frac{3}{8} - \frac{1}{6} = \frac{3 \times 6}{8 \times 6} - \frac{1 \times 8}{6 \times 8} = \frac{18}{48} - \frac{8}{48} = \frac{18-8}{48} = \frac{10}{48} = \frac{10 \div 2}{48 \div 2} = \frac{5}{24}$

10) $\frac{9}{21} - \frac{2}{7} = \frac{9}{21} - \frac{2 \times 3}{7 \times 3} = \frac{9}{21} - \frac{6}{21} = \frac{9-6}{21} = \frac{3}{21} = \frac{3 \div 3}{21 \div 3} = \frac{1}{7}$

11) $\frac{3}{4} \div \frac{9}{12} = \frac{3}{4} \times \frac{12}{9} = \frac{36}{36} = 1$

12) $\frac{7}{10} \div \frac{21}{20} = \frac{7}{10} \times \frac{20}{21} = \frac{140}{210} = \frac{140 \div 70}{210 \div 70} = \frac{2}{3}$

13) $\frac{12}{21} \div \frac{3}{7} = \frac{12 \div 3}{21 \div 3} \times \frac{7}{3} = \frac{4}{7} \times \frac{7}{3} = \frac{28}{21} = \frac{28 \div 7}{21 \div 7} = \frac{4}{3}$

14) $\frac{12}{5} \times \frac{10}{24} = \frac{120}{120} = 1$

15) $\frac{33}{36} \times \frac{3}{4} = \frac{33 \div 3}{36 \div 3} \times \frac{3}{4} = \frac{11}{12} \times \frac{3}{4} = \frac{33 \div 3}{48 \div 3} = \frac{11}{16}$

16) $\frac{7}{9} \times \frac{1}{3} = \frac{7}{27}$

17) $3\frac{1}{2} + 1\frac{3}{4} \rightarrow 3 + \frac{1}{2} + 1 + \frac{3}{4} \rightarrow 3 + 1 = 4, \quad \frac{1}{2} + \frac{3}{4} = \frac{1 \times 2}{2 \times 2} + \frac{3}{4} = \frac{2}{4} + \frac{3}{4} = \frac{2+3}{4} = \rightarrow$

$\frac{5}{4} = 1\frac{1}{4}, 4 + 1\frac{1}{4} = 5\frac{1}{4}$

18) $4\frac{1}{8} + 2\frac{7}{8} \rightarrow 4 + \frac{1}{8} + 2 + \frac{7}{8} \rightarrow 4 + 2 = 6, \quad \frac{1}{8} + \frac{7}{8} = \frac{1+7}{8} = \frac{8}{8} = 1 \rightarrow 6 + 1 = 7$

19) $4\frac{1}{2} + 2\frac{3}{8} \rightarrow 4 + \frac{1}{2} + 2 + \frac{3}{8} \rightarrow 4 + 2 = 6, \quad \frac{1}{2} + \frac{3}{8} = \frac{1 \times 4}{2 \times 4} + \frac{3}{8} = \frac{4}{8} + \frac{3}{8} = \frac{4+3}{8} = \frac{7}{8} \rightarrow$

$6 + \frac{7}{8} = 6\frac{7}{8}$

20) $1\frac{2}{21} + 4\frac{4}{7} \rightarrow 1 + \frac{2}{21} + 4 + \frac{4}{7} \rightarrow 1 + 4 = 5, \quad \frac{2}{21} + \frac{4}{7} = \frac{2}{21} + \frac{4 \times 3}{7 \times 3} + = \frac{2}{21} + \frac{12}{21} = \frac{2+12}{21} = \frac{14}{21} =$

$\frac{14 \div 7}{21 \div 7} = \frac{2}{3}$

$\rightarrow 5 + \frac{2}{3} = 5\frac{2}{3}$

21) $6\frac{3}{5} + 1\frac{2}{3} \rightarrow 6 + \frac{3}{5} + 1 + \frac{2}{3} \rightarrow 6 + 1 = 7, \quad \frac{3}{5} + \frac{2}{3} = \frac{3 \times 3}{5 \times 3} + \frac{2 \times 5}{3 \times 5} = \frac{9}{15} + \frac{10}{15} = \frac{9+10}{15} = \frac{19}{15} = 1\frac{4}{15} \rightarrow$

$$7 + 1\frac{4}{15} = 8\frac{4}{15}$$

22) $2\frac{3}{11} + 3\frac{1}{2} \rightarrow 2 + \frac{3}{11} + 3 + \frac{1}{2} \rightarrow 2 + 3 = 5, \frac{3}{11} + \frac{1}{2} = \frac{3\times2}{11\times2} + \frac{1\times11}{2\times11} = \frac{6}{22} + \frac{11}{22} = \frac{6+11}{22} = \frac{17}{22}$

$\rightarrow 5 + \frac{17}{22} = 5\frac{17}{22}$

23) $6\frac{2}{3} - 4\frac{1}{3} \rightarrow 6 + \frac{2}{3} - 4 - \frac{1}{3} \rightarrow 6 - 4 = 2, \frac{2}{3} - \frac{1}{3} = \rightarrow \frac{2-1}{3} = \frac{1}{3} \rightarrow 2 + \frac{1}{3} = 2\frac{1}{3}$

24) $5\frac{2}{5} - 3\frac{1}{5} \rightarrow 5 + \frac{2}{5} - 3 - \frac{1}{5} \rightarrow 5 - 3 = 2, \frac{2}{5} - \frac{1}{5} = \frac{1}{5} \rightarrow 2 + \frac{1}{5} = 2\frac{1}{5}$

25) $8\frac{1}{2} - 3\frac{1}{4} \rightarrow 8 + \frac{1}{2} - 3 - \frac{1}{4} \rightarrow 8 - 3 = 5, \frac{1}{2} - \frac{1}{4} = \frac{1\times2}{2\times2} - \frac{1}{4} = \frac{1}{4} \rightarrow 5 + \frac{1}{4} = 5\frac{1}{4}$

26) $7\frac{2}{3} - 2\frac{1}{6} \rightarrow 7 + \frac{2}{3} - 2 - \frac{1}{6} \rightarrow 7 - 2 = 5, \frac{2}{3} - \frac{1}{6} = \frac{2\times2}{3\times2} - \frac{1}{6} = \frac{3\div3}{6\div3} = \frac{1}{2} \rightarrow 5 + \frac{1}{2} = 5\frac{1}{2}$

27) $1\frac{2}{3} \times 2\frac{3}{4} \rightarrow 1\frac{2}{3} = \frac{1\times3+2}{3} = \frac{5}{3}, 2\frac{3}{4} = \frac{2\times4+3}{4} = \frac{11}{4} \rightarrow \frac{5}{3} \times \frac{11}{4} = \frac{5\times11}{3\times4} = \frac{55}{12} = 4\frac{7}{12}$

28) $1\frac{1}{6} \times 1\frac{3}{5} \rightarrow 1\frac{1}{6} = \frac{1\times6+1}{6} = \frac{7}{6}, 1\frac{3}{5} = \frac{1\times5+3}{5} = \frac{8}{5} \rightarrow \frac{7}{6} \times \frac{8}{5} = \frac{7\times8}{6\times5} = \frac{56}{30} = \frac{56\div2}{30\div2} = \frac{28}{15} = 1\frac{13}{15}$

29) $4\frac{1}{2} \times 1\frac{2}{3} \rightarrow 4\frac{1}{2} = \frac{4\times2+1}{2} = \frac{9}{2}, 1\frac{2}{3} = \frac{1\times3+2}{3} = \frac{5}{3} \rightarrow \frac{9}{2} \times \frac{5}{3} = \frac{9\times5}{2\times3} = \frac{45}{6} = \frac{45\div3}{6\div3} = \frac{15}{2} = 7\frac{1}{2}$

30) $2\frac{1}{2} \times 4\frac{4}{5} \rightarrow 2\frac{1}{2} = \frac{2\times2+1}{2} = \frac{5}{2}, 4\frac{4}{5} = \frac{4\times5+4}{5} = \frac{24}{5} \rightarrow \frac{5}{2} \times \frac{24}{5} = \frac{5\times24}{2\times5} = \frac{120}{10} = 12$

31) $2\frac{1}{5} \times 4\frac{1}{2} \rightarrow 2\frac{1}{5} = \frac{2\times5+1}{5} = \frac{11}{5}, 4\frac{1}{2} = \frac{4\times2+1}{2} = \frac{9}{2} \rightarrow \frac{11}{5} \times \frac{9}{2} = \frac{11\times9}{5\times2} = \frac{99}{10} = 9\frac{9}{10}$

32) $1\frac{1}{9} \times 2\frac{3}{5} \rightarrow 1\frac{1}{9} = \frac{1\times9+1}{9} = \frac{10}{9}, 2\frac{3}{5} = \frac{2\times5+3}{5} = \frac{13}{5} \rightarrow \frac{10}{9} \times \frac{13}{5} = \frac{10\times13}{9\times5} = \frac{130}{45} = \frac{130\div5}{45\div5} \rightarrow$

$\frac{26}{9} = 2\frac{8}{9}$

33) $3\frac{1}{3} \div 1\frac{2}{3} \rightarrow 3\frac{1}{3} = \frac{3\times3+1}{3} = \frac{10}{3}, 1\frac{2}{3} = \frac{1\times3+2}{3} = \frac{5}{3} \rightarrow \frac{10}{3} \div \frac{5}{3} = \frac{10}{3} \times \frac{3}{5} = \frac{30}{15} = 2$

34) $4\frac{2}{3} \div 2\frac{1}{2} \rightarrow 4\frac{2}{3} = \frac{4\times3+2}{3} = \frac{14}{3}, 2\frac{1}{2} = \frac{2\times2+1}{2} = \frac{5}{2} \rightarrow \frac{14}{3} \div \frac{5}{2} = \frac{14}{3} \times \frac{2}{5} = \frac{28}{15} = 1\frac{13}{15}$

35) $6\frac{1}{5} \div 2\frac{1}{3} \rightarrow 6\frac{1}{5} = \frac{6\times5+1}{5} = \frac{31}{5}, 2\frac{1}{3} = \frac{2\times3+1}{3} = \frac{7}{3} \rightarrow \frac{31}{5} \div \frac{7}{3} = \frac{31}{5} \times \frac{3}{7} = \frac{93}{35} = 2\frac{23}{35}$

36) $2\frac{2}{3} \div 1\frac{4}{9} \rightarrow 2\frac{2}{3} = \frac{2\times3+2}{3} = \frac{8}{3}, 1\frac{4}{9} = \frac{1\times9+4}{9} = \frac{13}{9} \rightarrow \frac{8}{3} \div \frac{13}{9} = \frac{8}{3} \times \frac{9}{13} = \frac{72\div3}{39\div3} = \frac{24}{13} = 1\frac{11}{13}$

37) $4\frac{1}{6} \div 2\frac{1}{8} \rightarrow 4\frac{1}{6} = \frac{4\times6+1}{6} = \frac{25}{6}, 2\frac{1}{8} = \frac{2\times8+1}{8} = \frac{17}{8} \rightarrow \frac{25}{6} \div \frac{17}{8} = \frac{25}{6} \times \frac{8}{17} = \frac{200\div2}{102\div2} \rightarrow$

$\frac{100}{51} = 1\frac{49}{51}$

38) $3\frac{2}{5} \div 1\frac{5}{4} \to 3\frac{2}{5} = \frac{3 \times 5 + 2}{5} = \frac{17}{5}, 1\frac{5}{4} = \frac{1 \times 4 + 5}{4} = \frac{9}{4} \to \frac{17}{5} \div \frac{9}{4} = \frac{17}{5} \times \frac{4}{9} = \frac{68}{45} = 1\frac{23}{45}$

39) David ate $\frac{1}{3}$ of 6 parts of his pizza→ $\frac{1}{3} \times 6 = \frac{6 \div 3}{3 \div 3} = 2$ →It means 2 parts out of 6 parts and left 4 parts. Sara ate $\frac{1}{2}$ of 6 parts of her pizza:

→ $\frac{1}{2} \times 6 = \frac{6 \div 2}{2 \div 2} = 3$ →It means 3 parts out of 6 parts and left 3 parts. Therefore, they ate $(2 + 3)$ parts out of $(6 + 6)$ parts of their pizza and left $(4 + 3)$ parts out of $(6 + 6)$ parts of their pizza that equals to: $\frac{7}{12}$

40) Jake run $9\frac{1}{3}$ miles on Saturday and $2 \times \left(9\frac{1}{3}\right)$ miles on Monday and Wednesday. Jake wants to run a total of 50 miles this week. Therefore, $9\frac{1}{3} + 2 \times \left(9\frac{1}{3}\right)$ should be subtracted from 50:

$50 - \left(9\frac{1}{3} + \left(2 \times 9\frac{1}{3}\right)\right) = 50 - \left(\frac{9 \times 3 + 1}{3} + \left(2 \times \frac{9 \times 3 + 1}{3}\right)\right) = 50 - \left(\frac{28}{3} + \frac{56}{3}\right) = 50 - \left(\frac{28 + 56}{3}\right) = 50 - \left(\frac{84}{3}\right) = 50 - (28) = 22$ miles.

41) Four times of 21,000 is 84,000. One third of them cancelled their tickets. One third of 84,000 Equals 28,000. $(\frac{1}{3} \times 84,000 = 28,000)$.

84,000 – 28,000 = 56,000 fans are attending this week.

42) Let x be the total number of balls. Then: $\frac{1}{2}x + \frac{1}{4}x + \frac{1}{8}x + 16 = x$

$\left(\frac{1}{2} + \frac{1}{4} + \frac{1}{8}\right)x + 16 = x \to \left(\frac{1 \times 4}{2 \times 4} + \frac{1 \times 2}{4 \times 2} + \frac{1}{8}\right)x + 16 = x \to$

$\left(\frac{4}{8} + \frac{2}{8} + \frac{1}{8}\right)x + 16 = x \to \left(\frac{7}{8}\right)x + 16 = x \to 16 = x - \frac{7}{8}x \to 16 = \frac{1}{8}x$

→ Multiply both sides by 8: $16 \times 8 = \frac{1}{8}x \times 8 \to 128 = x$

x is the total number of balls. Therefore, number of white balls is:

$\frac{1}{4}x = \frac{1}{4} \times 128 = 32$

 **DAY 2** # Decimals and Integers

Math topics that you'll learn in this chapter:

1. Comparing Decimals
2. Rounding Decimals
3. Adding and Subtracting Decimals
4. Multiplying and Dividing Decimals
5. Adding and subtracting Integers
6. Multiplying and Dividing Integers
7. Order of Operations
8. Integers and Absolute Value

15

Comparing Decimals

☆ A decimal is a fraction written in a special form. For example, instead of writing $\frac{1}{2}$ you can write: 0.5

☆ A Decimal Number contains a Decimal Point. It separates the whole number part from the fractional part of a decimal number.

☆ Let's review decimal place values: Example: 45.3861

4: tens	5: ones	3: tenths
8: hundredths	6: thousandths	1: tens thousandths

☆ To compare two decimals, compare each digit of two decimals in the same place value. Start from left. Compare hundreds, tens, ones, tenth, hundredth, etc.

☆ To compare numbers, use these symbols:

Equal to =	Less than <	Greater than >
Less than or equal ≤	Greater than or equal ≥

Examples:

Example 1. Compare 0.05 and 0.50.

Solution: 0.50 is greater than 0.05, because the tenth place of 0.50 is 5, but the tenth place of 0.05 is zero. Then: 0.05 < 0.50

Example 2. Compare 0.0512 and 0.181.

Solution: 0.181 is greater than 0.0512, because the tenth place of 0.181 is 1, but the tenth place of 0.0512 is zero. Then: 0.0512 < 0.181

Rounding Decimals

☆ We can round decimals to a certain accuracy or number of decimal places. This is used to make calculations easier to do and results easier to understand when exact values are not too important.

☆ First, you'll need to remember your place values: For example: 12.4869

    1: tens   2: ones   4: tenths

    8: hundredths   6: thousandths   9: tens thousandths

☆ To round a decimal, first find the place value you'll round to.

☆ Find the digit to the right of the place value you're rounding to. If it is 5 or bigger, add 1 to the place value you're rounding to and remove all digits on its right side. If the digit to the right of the place value is less than 5, keep the place value and remove all digits on the right.

Examples:

Example 1. Round 3.2568 to the thousandth-place value.

Solution: First, look at the next place value to the right, (tens thousandths). It's 8 and it is greater than 5. Thus add 1 to the digit in the thousandth place. The thousandth place is 6. $\rightarrow 6 + 1 = 7$, then, the answer is 3.257.

Example 2. Round 2.3628 to the nearest hundredth.

Solution: First, look at the digit to the right of hundredth (thousandths place value). It's 2 and it is less than 5, thus remove all the digits to the right of hundredth place. Then, the answer is 2.36.

Adding and Subtracting Decimals

☆ Line up the decimal numbers.

☆ Add zeros to have the same number of digits for both numbers if necessary.

☆ Remember your place values: For example: 73.5196

7: tens	3: ones	5: tenths
1: hundredths	9: thousandths	6: tens thousandths

☆ Add or subtract using column addition or subtraction.

Examples:

Example 1. Add. $2.6 + 3.25 =$

Solution: First, line up the numbers: $\begin{array}{r} 2.6 \\ + 3.25 \\ \hline \end{array}$ →Add a zero to have the same number of digits for both numbers. $\begin{array}{r} 2.60 \\ + 3.25 \\ \hline \end{array}$ →Start with the hundredths place: $0 + 5 = 5$, $\begin{array}{r} 2.60 \\ + 3.25 \\ \hline 5 \end{array}$ →Continue with tenths place: $6 + 2 = 8$, $\begin{array}{r} 2.60 \\ + 3.25 \\ \hline .85 \end{array}$ →Add the ones place: $2 + 3 = 5$, $\begin{array}{r} 2.60 \\ + 3.25 \\ \hline 5.85 \end{array}$. The answer is 5.85.

Example 2. Find the difference. $4.26 - 3.12 =$

Solution: First, line up the numbers: $\begin{array}{r} 4.26 \\ - 3.12 \\ \hline \end{array}$ →Start with the hundredths place: $6 - 2 = 4$, $\begin{array}{r} 4.26 \\ - 3.12 \\ \hline 4 \end{array}$ → Continue with tenths place. $2 - 1 = 1$, $\begin{array}{r} 4.26 \\ - 3.12 \\ \hline .14 \end{array}$ →Subtract the ones place. $4 - 3 = 1$, $\begin{array}{r} 4.26 \\ - 3.12 \\ \hline 1.14 \end{array}$

Multiplying and Dividing Decimals

For multiplying decimals:

✪ Ignore the decimal point and set up and multiply the numbers as you do with whole numbers.

✪ Count the total number of decimal places in both of the factors.

✪ Place the decimal point in the product.

For dividing decimals:

✪ If the divisor is not a whole number, move the decimal point to the right to make it a whole number. Do the same for the dividend.

✪ Divide similar to whole numbers.

Examples:

Example 1. Find the product. $0.53 \times 0.32 =$

Solution: Set up and multiply the numbers as you do with whole numbers. Line up the numbers: $\begin{array}{r} 53 \\ \times 32 \end{array}$ →Start with the ones place then continue with other digits $\rightarrow \begin{array}{r} 53 \\ \times 32 \\ \hline 1,696 \end{array}$. Count the total number of decimal places in both of the factors. There are four decimals digits. (two for each factor 0.53 and 0.32) Then: $0.53 \times 0.32 = 0.1696$

Example 2. Find the quotient. $1.50 \div 0.5 =$

Solution: The divisor is not a whole number. Multiply it by 10 to get 5:
$$\rightarrow 0.5 \times 10 = 5$$
Do the same for the dividend to get 15.→ $1.50 \times 10 = 15$
Now, divide $15 \div 5 = 3$. The answer is 3.

Adding and Subtracting Integers

✩ Integers include zero, counting numbers, and the negative of the counting numbers $\{\ldots, -3, -2, -1, 0, 1, 2, 3, \ldots\}$

✩ Add a positive integer by moving to the right on the number line. (you will get a bigger number)

✩ Add a negative integer by moving to the left on the number line. (you will get a smaller number)

✩ Subtract an integer by adding its opposite.

Examples:

Example 1. Solve. $(-3) - (-5) =$

Solution: Keep the first number and convert the sign of the second number to its opposite. (change subtraction into addition. Then: $(-3) + 5 = 2$

Example 2. Solve. $5 + (2 - 8) =$

Solution: First, subtract the numbers in brackets, $2 - 8 = -6$
Then: $5 + (-6) = \rightarrow$ change addition into subtraction: $5 - 6 = -1$

Example 3. Solve. $(8 - 15) + 14 =$

Solution: First, subtract the numbers in brackets, $8 - 15 = -7$
Then: $-7 + 14 = \rightarrow -7 + 14 = 7$

Example 4. Solve. $18 + (-5 - 17) =$

Solution: First, subtract the numbers in brackets, $-5 - 17 = -22$
Then: $18 + (-22) = \rightarrow$ change addition into subtraction: $18 - 22 = -4$

Multiplying and Dividing Integers

Use the following rules for multiplying and dividing integers:

☆ (negative) × (negative) = positive

☆ (negative) ÷ (negative) = positive

☆ (negative) × (positive) = negative

☆ (negative) ÷ (positive) = negative

☆ (positive) × (positive) = positive

☆ (positive) ÷ (negative) = negative

Examples:

Example 1. Solve. $5 \times (-2) =$

Solution: Use this rule: (positive) × (negative) = negative.
Then: $(5) \times (-2) = -10$

Example 2. Solve. $(-2) + (-30 \div 6) =$

Solution: First, divide -30 by 6, the numbers in brackets, use this rule:
(negative) ÷ (positive) = negative. Then: $-30 \div 6 = -5$
$(-2) + (-30 \div 6) = (-2) + (-5) = -2 - 5 = -7$

Example 3. Solve. $(13 - 16) \times (-3) =$

Solution: First, subtract the numbers in brackets,
$13 - 16 = -3 \rightarrow (-3) \times (-3) =$
Now use this rule: (negative) × (negative) = positive → $(-3) \times (-3) = 9$

Example 4. Solve. $(18 - 3) \div (-5) =$

Solution: First, subtract the numbers in brackets,
$18 - 3 = 15 \rightarrow (15) \div (-5) =$
Now use this rule: (positive) ÷ (negative) = negative → $(15) \div (-5) = -3$

Order of Operations

☆ In Mathematics, "operations" are addition, subtraction, multiplication, division, exponentiation (written as b^n) and grouping.

☆ When there is more than one math operation in an expression, use PEMDAS: (to memorize this rule, remember the phrase "Please Excuse My Dear Aunt Sally".)

❖ Parentheses

❖ Exponents

❖ Multiplication and Division (from left to right)

❖ Addition and Subtraction (from left to right)

Examples:

Example 1. Calculate. $(3 + 5) \div (8 \div 4) =$

Solution: First, simplify inside parentheses:
$(3 + 5) \div (8 \div 4) = (8) \div (8 \div 4) = (8) \div (2)$. Then: $(8) \div (2) = 4$

Example 2. Solve. $(4 \times 3) - (14 - 3) =$

Solution: First, calculate within parentheses: $(4 \times 3) - (14 - 3) = (12) - (11)$, Then: $(12) - (11) = 1$

Example 3. Calculate. $-3[(6 \times 5) \div (5 \times 3)] =$

Solution: First, calculate within parentheses:
$-3[(6 \times 5) \div (5 \times 3)] = -3[(30) \div (5 \times 3)] = -3[(30) \div (15)] = -3[2]$
Multiply -3 and 2. Then: $-3[2] = -6$

Example 4. Solve. $(32 \div 4) + (-25 + 5) =$

Solution: First, calculate within parentheses:
$(32 \div 4) + (-25 + 5) = (8) + (-20)$. Then: $(8) - (20) = -12$

Integers and Absolute Value

☆ The absolute value of a number is its distance from zero, in either direction, on the number line. For example, the distance of 9 and −9 from zero on number line is 9.

☆ The absolute value of an integer is the numerical value without its sign. (negative or positive)

☆ The vertical bar is used for absolute value as in $|x|$.

☆ The absolute value of a number is never negative; because it only shows, "how far the number is from zero".

Examples:

Example 1. Calculate. $|15 − 3| \times 6 =$

Solution: First, solve $|15 − 3| \rightarrow |15 − 3| = |12|$, the absolute value of 12 is 12, $|12| = 12$. Then: $12 \times 6 = 72$

Example 2. Solve. $|−35| \times |6 − 10| =$

Solution: First, find $|−35| \rightarrow$ the absolute value of −35 is 35. Then: $|−35| = 35$, $|−35| \times |6 − 10| =$
Now, calculate $|6 − 10| \rightarrow |6 − 10| = |−4|$, the absolute value of −4 is 4. $|−4| = 4$
Then: $35 \times 4 = 140$

Example 3. Solve. $|12 − 6| \times \frac{|−4 \times 5|}{3} =$

Solution: First, calculate $|12 − 6| \rightarrow |12 − 6| = |6|$, the absolute value of 6 is 6, $|6| = 6$. Then: $6 \times \frac{|−4 \times 5|}{3} =$
Now calculate $|−4 \times 5| \rightarrow |−4 \times 5| = |−20|$, the absolute value of −20 is 20, $|−20| = 20$. Then: $6 \times \frac{20}{3} = \frac{6 \times 20}{3} = \frac{120}{3} = 40$

Day 2: Practices

🖎 Compare. Use >, =, and <

1) $0.3 \ \square \ 0.2$

2) $0.98 \ \square \ 0.71$

3) $5.01 \ \square \ 5.0100$

4) $0.427 \ \square \ 0.435$

🖎 Round each decimal to the nearest whole number.

5) 4.9

6) 6.3

7) 75.66

8) 93.03

🖎 Find the sum or difference.

9) $2.5 + 11.1 =$

10) $12.83 + 14.11 =$

11) $13.8 - 9.2 =$

12) $43.55 - 21.32 =$

🖎 Find the product or quotient.

13) $5.1 \times 0.2 =$

14) $0.35 \times 1.2 =$

15) $2.1 \div 0.3$

16) $25.5 \div 0.5$

🖎 Find each sum or difference.

17) $-6 + 17 =$

18) $12 - 21 =$

19) $31 - (-4) =$

20) $(7 + 5) + (8 - 3) =$

21) $(2 - 3) - (15 - 11) =$

22) $(-8 - 7) - (-6 - 2) =$

✎ Solve.

23) $2 \times (-4) =$

24) $(-5) \times (-3) =$

25) $(-15) \div 5 =$

26) $(-4) \times (-5) \times (-2) =$

27) $(-6 + 36) \div (-2) =$

28) $(-25 + 5) \times (-5 - 3) =$

✎ Evaluate each expression.

29) $5 - (2 \times 3) =$

30) $(6 \times 5) - 8 =$

31) $(-5 \times 3) + 4 =$

32) $(-35 \div 5) - (12 + 2) =$

33) $4 \times [(2 \times 3) \div (-3 + 1)] =$

34) $35 \div [(6 - 1) \times (7 - 8)] =$

✎ Find the answers.

35) $|-3| + |7 - 9| =$

36) $|8 - 10| + |6 - 7| =$

37) $|-6 + 10| - |-9 - 3| =$

38) $3 + |2 - 1| + |3 - 12| =$

39) $-8 - |3 - 6| + |2 + 3| =$

40) $|-6| \times |-5.4| =$

41) $|3 \times (-4)| \times \frac{8}{3} =$

42) $|(-2) \times (-2)| \times \frac{1}{4} =$

43) $|-8| + |(-8) \times 2| =$

44) $|(-3) \times (-5)| \times |(-3) \times (-4)| =$

✎ Find the answers.

45) Round 4.2873 to the thousandth-place value

46) $[6 \times (-16) + 8] - (-4) + [4 \times 5] \div 2 =$

Day 2: Answers

1) $0.3 > 0.2$

2) $0.98 > 0.71$

3) $5.01 = 5.0100$

4) $0.427 < 0.435$

5) $4.9 \approx 5$

6) $6.3 \approx 6$

7) $75.66 \approx 76$

8) $93.03 \approx 93$

9) $\begin{matrix} 2.5 \\ +11.1 \\ \hline \end{matrix} \to 5+1=6 \to \begin{matrix} 2.5 \\ +11.1 \\ \hline .6 \end{matrix} \to 2+1=3 \to \begin{matrix} 2.5 \\ +11.1 \\ \hline 3.6 \end{matrix} \to 0+1=1 \to \begin{matrix} 2.5 \\ +11.1 \\ \hline 13.6 \end{matrix}$

10) $\begin{matrix} 12.83 \\ +14.11 \\ \hline \end{matrix} \to 3+1=4 \to \begin{matrix} 12.83 \\ +14.11 \\ \hline 4 \end{matrix} \to 8+1=9 \to \begin{matrix} 12.83 \\ +14.11 \\ \hline .94 \end{matrix} \to 2+4=6 \to \begin{matrix} 12.83 \\ +14.11 \\ \hline 6.94 \end{matrix} \to 1+1=2 \to \begin{matrix} 12.83 \\ +14.11 \\ \hline 26.94 \end{matrix}$

11) $\begin{matrix} 13.8 \\ -9.2 \\ \hline \end{matrix} \to 8-2=6 \to \begin{matrix} 13.8 \\ -9.2 \\ \hline .6 \end{matrix} \to 13-9=4 \to \begin{matrix} 13.8 \\ -9.2 \\ \hline 4.6 \end{matrix}$

12) $\begin{matrix} 43.55 \\ -21.32 \\ \hline \end{matrix} \to 5-2=3 \to \begin{matrix} 43.55 \\ -21.32 \\ \hline 3 \end{matrix} \to 5-3=2 \to \begin{matrix} 43.55 \\ -21.32 \\ \hline 23 \end{matrix} \to 3-1=2 \to \begin{matrix} 43.55 \\ -21.32 \\ \hline 2.23 \end{matrix} \to 4-2=2 \to \begin{matrix} 43.55 \\ -21.32 \\ \hline 22.23 \end{matrix}$

13) $5.1 \times 0.2 \to \begin{matrix} 51 \\ \times 2 \\ \hline 102 \end{matrix} \to 5.1 \times 0.2 = 1.02$

14) $0.35 \times 1.2 \to \begin{matrix} 35 \\ \times 12 \\ \hline 70 \\ +350 \\ \hline 420 \end{matrix} \to 0.35 \times 1.2 = 0.420$

15) $2.1 \div 0.3 \to \frac{2.1 \times 10}{0.3 \times 10} = \frac{21}{3} = 7$

16) $25.5 \div 0.5 \to \frac{25.50 \times 10}{0.5 \times 10} = \frac{255}{5} = 51$

17) $-6 + 17 = 17 - 6 = 11$

18) $12 - 21 = -9$

19) $31 - (-4) = 31 + 4 = 35$

20) $(7 + 5) + (8 - 3) = 12 + 5 = 17$

21) $(2 - 3) - (15 - 11) = (-1) - (4) = -1 - 4 = -5$

22) $(-8 - 7) - (-6 - 2) = (-15) - (-8) = -15 + 8 = -7$

23) Use this rule: (positive) $\times$ (negative) = negative $\to 2 \times (-4) = -8$

24) Use this rule: (negative) $\times$ (negative) = positive $\to (-5) \times (-3) = 15$

25) Use this rule: (negative) $\div$ (positive) = negative $\to (-15) \div 5 = -3$

26) Use these rules: [(negative) $\times$ (negative) = positive] and [(positive) $\times$ (negative) = negative]$\to (-4) \times (-5) \times (-2) = (20) \times (-2) = -40$

27) Use this rule: (positive)$\div$ (negative) = negative$\to$
    $(-6 + 36) \div (-2) = (30) \div (-2) = -15$

28) Use this rule: (negative) $\times$ (negative) = positive$\to$
    $(-25 + 5) \times (-5 - 3) = (-20) \times (-8) = -160$

29) $5 - (2 \times 3) = 5 - 6 = -1$

30) $(6 \times 5) - 8 = 30 - 8 = 22$

31) Use this rule: (negative) $\times$ (positive) = negative $\to (-5 \times 3) + 4 = -15 + 4 = -11$

32) Use this rule: (negative) $\div$ (positive) = negative $\to$
    $(-35 \div 5) - (12 + 2) = -7 - 14 = -21$

33) Use these rules: [(positive) × (negative) = negative] and [(positive) ÷ (negative) = negative]→ $4 \times [(2 \times 3) \div (-3 + 1)] = 4 \times [6 \div (-2)] = 4 \times (-3) = -12$

34) Use these rules: [(positive) × (negative) = negative] and [(positive) ÷ (negative) = negative]→ $35 \div [(6 - 1) \times (7 - 8)] = 35 \div [5 \times (-1)] = 35 \div (-5) = -7$

35) $|-3| + |7 - 9| = 3 + |-2| = 3 + 2 = 5$

36) $|8 - 10| + |6 - 7| = |-2| + |-1| = 2 + 1 = 3$

37) $|-6 + 10| - |-9 - 3| = |4| - |-12| = 4 - (12) = 4 - 12 = -8$

38) $3 + |2 - 1| + |3 - 12| = 3 + |1| + |-9| = 3 + 1 + (9) = 3 + 1 + 9 = 13$

39) $-8 - |3 - 6| + |2 + 3| = -8 - |-3| + |5| = -8 - (3) + 5 = -8 - 3 + 5 = -11 + 5 = -6$

40) $|-6| \times |-5.4| = 6 \times 5.4 = 32.4$

41) $|3 \times (-4)| \times \frac{8}{3} = |-12| \times \frac{8}{3} = 12 \times \frac{8}{3} = \frac{12 \times 8}{3} = 32$

42) $|(-2) \times (-2)| \times \frac{1}{4} = |4| \times \frac{1}{4} = 4 \times \frac{1}{4} = 1$

43) $|-8| + |(-8) \times 2| = 8 + |-16| = 8 + 16 = 24$

44) $|(-3) \times (-5)| \times |(-3) \times (-4)| = |15| \times |12| = 15 \times 12 = 180$

45) $4.2873 \approx 4.287$

46) $[6 \times (-16) + 8] - (-4) + [4 \times 5] \div 2 = [(-96) + 8] - (-4) + [4 \times 5] \div 2 = (-88) - (-4) + (20) \div 2 = (-88) - (-4) + 10 = (-88) + 4 + (10) = -84 + 10 = -74$

3 Ratios, Proportions and Percent

Math topics that you'll learn in this chapter:

1. Simplifying Ratios
2. Proportional Ratios
3. Similarity and Ratios
4. Percent Problems
5. Percent of increase and Decrease
6. Discount, Tax and Tip
7. Simple Interest

29

Simplifying Ratios

☆ Ratios are used to make comparisons between two numbers.

☆ Ratios can be written as a fraction, using the word "to", or with a colon.
Example: $\frac{3}{4}$ or "3 to 4" or 3:4

☆ You can calculate equivalent ratios by multiplying or dividing both sides
of the ratio by the same number.

Examples:

Example 1. Simplify. $10:2 =$

Solution: Both numbers 10 and 2 are divisible by $2 \Rightarrow 10 \div 2 = 5, 2 \div 2 = 1$.
Then: $10:2 = 5:1$

Example 2. Simplify. $\frac{6}{33} =$

Solution: Both numbers 6 and 33 are divisible by $3 \Rightarrow 33 \div 3 = 11, 6 \div 3 = 2$.
Then: $\frac{6}{33} = \frac{2}{11}$

Example 3. There are 30 students in a class and 12 are girls. Find the ratio of
girls to boys in that class.

Solution: Subtract 12 from 30 to find the number of boys in the class.
$30 - 12 = 18$. There are 18 boys in the class. So, the ratio of girls to boys is $12:18$.
Now, simplify this ratio. Both 18 and 12 are divisible by 6.
Then: $18 \div 6 = 3$, and $12 \div 6 = 2$. In the simplest form, this ratio is $2:3$

Example 4. A recipe calls for butter and sugar in the ratio $2:3$. If you're using 6
cups of butter, how many cups of sugar should you use?

Solution: Since you use 6 cups of butter, or 3 times as much, you need to
multiply the amount of sugar by 3. Then: $3 \times 3 = 9$. So, you need to use 9 cups of
sugar. You can solve this using equivalent fractions: $\frac{2}{3} = \frac{6}{9}$

Proportional Ratios

☆ Two ratios are proportional if they represent the same relationship.

☆ A proportion means that two ratios are equal. It can be written in two ways: $\frac{a}{b} = \frac{c}{d}$ $a : b = c : d$

☆ The proportion $\frac{a}{b} = \frac{c}{d}$ can be written as: $a \times d = c \times b$

Examples:

Example 1. Solve this proportion for x. $\frac{3}{4} = \frac{9}{x}$

Solution: Use cross multiplication: $\frac{3}{4} = \frac{9}{x} \Rightarrow 3 \times x = 4 \times 9 \Rightarrow 3x = 36$
Divide both sides by 3 to find x: $x = \frac{36}{3} \Rightarrow x = 12$

Example 2. If a box contains red and blue balls in ratio of $4 : 7$ red to blue, how many red balls are there if 49 blue balls are in the box?

Solution: Write a proportion and solve. $\frac{4}{7} = \frac{x}{49}$
Use cross multiplication: $4 \times 49 = 7 \times x \Rightarrow 196 = 7x$
Divide to find x: $x = \frac{196}{7} \Rightarrow x = 28$. There are 28 red balls in the box.

Example 3. Solve this proportion for x. $\frac{5}{8} = \frac{20}{x}$

Solution: Use cross multiplication: $\frac{5}{8} = \frac{20}{x} \Rightarrow 5 \times x = 8 \times 20 \Rightarrow 5x = 160$
Divide to find x: $x = \frac{160}{5} \Rightarrow x = 32$

Example 4. Solve this proportion for x. $\frac{7}{9} = \frac{21}{x}$

Solution: Use cross multiplication: $\frac{7}{9} = \frac{21}{x} \Rightarrow 7 \times x = 9 \times 21 \Rightarrow 7x = 189$
Divide to find x: $x = \frac{189}{7} \Rightarrow x = 27$

Similarity and Ratios

☆ Two figures are similar if they have the same shape.

☆ Two or more figures are similar if the corresponding angles are equal, and the corresponding sides are in proportion.

Examples:

Example 1. The following triangles are similar. What is the value of the unknown side?

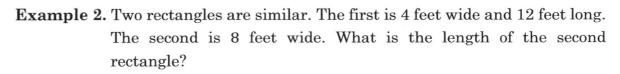

Solution: Find the corresponding sides and write a proportion.

$\frac{9}{18} = \frac{8}{x}$. Now, use the cross product to solve for x:

$\frac{9}{18} = \frac{8}{x} \rightarrow 9 \times x = 18 \times 8 \rightarrow 9x = 144$. Divide both sides by 9. Then: $9x = 144 \rightarrow x = \frac{144}{9} \rightarrow x = 16$

The missing side is 16.

Example 2. Two rectangles are similar. The first is 4 feet wide and 12 feet long. The second is 8 feet wide. What is the length of the second rectangle?

Solution: Let's put x for the length of the second rectangle. Since two rectangles are similar, their corresponding sides are in proportion. Write a proportion and solve for the missing number.

$\frac{4}{8} = \frac{12}{x} \rightarrow 4x = 8 \times 12 \rightarrow 4x = 96 \rightarrow x = \frac{96}{4} = 24$

The length of the second rectangle is 24 feet.

Percent Problems

★ Percent is a ratio of a number and 100. It always has the same denominator, 100. The percent symbol is "%".

★ Percent means "per 100". So, 20% is $\frac{20}{100}$.

★ In each percent problem, we are looking for the base, or the part or the percent.

★ Use these equations to find each missing section in a percent problem:

  ❖ Base = Part ÷ Percent

  ❖ Part = Percent × Base

  ❖ Percent = Part ÷ Base

Examples:

Example 1. What is 30% of 60?

Solution: In this problem, we have the percent (30%) and the base (60) and we are looking for the "part". Use this formula: $Part = Percent \times Base$.
Then: $Part = 30\% \times 60 = \frac{30}{100} \times 60 = 0.30 \times 60 = 18$. The answer: 30% of 60 is 18.

Example 2. 20 is what percent of 400?

Solution: In this problem, we are looking for the percent. Use this equation:
$Percent = Part \div Base \rightarrow Percent = 20 \div 400 = 0.05 = 5\%$.
Then: 20 is 5 percent of 400.

Example 3. 70 is 25 percent of what number?

Solution: In this problem, we are looking for the base. Use this equation:
$Base = Part \div Percent \rightarrow Base = 70 \div 25\% = 70 \div 0.25 = 280$
Then: 70 is 25 percent of 280.

Percent of Increase and Decrease

☆ Percent of change (increase or decrease) is a mathematical concept that represents the degree of change over time.

☆ To find the percentage of increase or decrease:

1. New Number–Original Number

2. (The result÷Original Number)× 100

☆ Or use this formula: Percent of change=$\frac{new\ number - original\ number}{original\ number} \times 100$

☆ Note: If your answer is a negative number, then this is a percentage decrease. If it is positive, then this is a percentage increase.

Examples:

Example 1. The price of a shirt increases from $40 to $44. What is the percentage increase?

Solution: First, find the difference: $44 - 40 = 4$

Then: $(4 \div 40) \times 100 = \frac{4}{40} \times 100 = 10$. The percentage increase is 10%. It means that the price of the shirt increased by 10%.

Example 2. The price of a table decreased from $50 to $25. What is the percent of decrease?

Solution: Use this formula:

$$Percent\ of\ change = \frac{new\ number - original\ number}{original\ number} \times 100 =$$

$\frac{25-50}{50} \times 100 = \frac{-25}{50} \times 100 = -50$. The percentage decrease is 50. (the negative sign means percentage decrease) Therefore, the price of the table decreased by 50%.

Discount, Tax and Tip

☆ To find the discount: Multiply the regular price by the rate of discount

☆ To find the selling price: Original price − discount

☆ To find tax: Multiply the tax rate to the taxable amount (income, property value, etc.)

☆ To find the tip, multiply the rate to the selling price.

Examples:

Example 1. With an 25% discount, Ella saved $50 on a dress. What was the original price of the dress?

Solution: let x be the original price of the dress. Then: 25 % of $x = 50$. Write an equation and solve for x: $0.25 \times x = 50 \rightarrow x = \frac{50}{0.25} = 200$. The original price of the dress was $200.

Example 2. Sophia purchased a new computer for a price of $820 at the Apple Store. What is the total amount her credit card is charged if the sales tax is 10%?

Solution: The taxable amount is $820, and the tax rate is 10%. Then:
$$Tax = 0.10 \times 820 = 82$$
Final price = Selling price + Tax → final price = $820 + $82 = $902

Example 3. Nicole and her friends went out to eat at a restaurant. If their bill was $50 and they gave their server a 12% tip, how much did they pay altogether?

Solution: First, find the tip. To find the tip, multiply the rate to the bill amount. $Tip = 50 \times 0.12 = 6$. The final amount is: $50 + $6 = $56

Simple Interest

☆ Simple Interest: The charge for borrowing money or the return for lending it.

☆ Simple interest is calculated on the initial amount (principal).

☆ To solve a simple interest problem, use this formula:

$$Interest = principal \times rate \times time \rightarrow (I = p \times r \times t = prt)$$

Examples:

Example 1. Find simple interest for $250 investment at 6% for 5 years.

Solution: Use Interest formula:
$I = prt$ ($P = \$250, r = 6\% = \frac{6}{100} = 0.06$ and $t = 5$)
Then: $I = 250 \times 0.06 \times 5 = \$75$

Example 2. Find simple interest for $1,300 at 3% for 2 years.

Solution: Use Interest formula:
$I = prt$ ($P = \$1,300, r = 3\% = \frac{3}{100} = 0.03$ and $t = 2$)
Then: $I = 1,300 \times 0.03 \times 2 = \$78.00$

Example 3. Andy received a student loan to pay for his educational expenses this year. What is the interest on the loan if he borrowed $5,100 at 5% for 4 years?

Solution: Use Interest formula: $I = prt$. $P = \$5,100$, r $= 5\% = 0.05$ and $t = 4$
Then: $I = 5,100 \times 0.05 \times 4 = \$1,020$

Example 4. Bob is starting his own small business. He borrowed $28,000 from the bank at an 7% rate for 6 months. Find the interest Bob will pay on this loan.

Solution: Use Interest formula:
$I = prt$. $P = \$28,000, r = 7\% = 0.07$ and $t = 0.5$ (6 months is half year). Then:
$I = 28,000 \times 0.07 \times 0.5 = \$980$

Day 3: Practices

✍ Reduce each ratio.

1) $3:15 = $ ___ : ___

2) $8:72 = $ ___ : ___

3) $15:25 = $ ___ : ___

4) $35:10 = $ ___ : ___

5) $36:42 = $ ___ : ___

6) $24:64 = $ ___ : ___

✍ Solve.

7) In Jack's class, 48 of the students are tall and 20 are short. In Michael's class 28 students are tall and 12 students are short. Which class has a higher ratio of tall to short students? _____

8) You can buy 7 cans of green beans at a supermarket for $7.49. How much does it cost to buy 21 cans of green beans? _____

✍ Solve each proportion.

9) $\frac{3}{4} = \frac{12}{x} \rightarrow x = $ _____

10) $\frac{2}{5} = \frac{x}{20} \rightarrow x = $ _____

11) $\frac{9}{x} = \frac{3}{5} \rightarrow x = $ _____

12) $\frac{x}{24} = \frac{5}{6} \rightarrow x = $ _____

13) $\frac{8}{4} = \frac{x}{16} \rightarrow x = $ _____

14) $\frac{3}{x} = \frac{12}{16} \rightarrow x = $ _____

15) $\frac{24}{32} = \frac{3}{x} \rightarrow x = $ _____

16) $\frac{x}{7} = \frac{21}{49} \rightarrow x = $ _____

✍ Solve each problem.

17) Two rectangles are similar. The first is 6 *feet* wide and 36 *feet* long. The second is 10 *feet* wide. What is the length of the second rectangle? _____

18) Two rectangles are similar. One is 4.6 *meters* by 7 *meters*. The longer side of the second rectangle is 28 *meters*. What is the other side of the second rectangle? _____

✒ Solve each problem.

19) What is 15% of 60? ____

20) What is 20% of 500? ____

21) 25 what is percent of 250? __

22) 30 is what percent of 150? ___

23) 15 is 10 percent of what number? ___

24) 25 is 5 percent of what number? ___

✒ Solve each problem.

25) Bob got a raise, and his hourly wage increased from $15 to $21. What is the percent increase? ____ %

26) A $45 shirt now selling for $36 is discounted by what percent? ____ %

✒ Find the selling price of each item.

27) Original price of a computer: $500

Tax: 5%, Selling price: $_____

28) Nicolas hired a moving company. The company charged $500 for its services, and Nicolas gives the movers a 14% tip. How much does Nicolas tip the movers? $____

29) Mason has lunch at a restaurant and the cost of his meal is $60. Mason wants to leave a 10% tip. What is Mason's total bill, including tip? $_____

✒ Determine the simple interest for the following loans.

30) $800 at 3% for 2 years. $__

31) $260 at 10% for 5 years. $__

32) $380 at 4% for 3 years. $__

33) $1,200 at 2% for 1 years. $__

Day 3: Answers

1) $3:15 \rightarrow 3 \div 3 = 1, \ 15 \div 3 = 5 \rightarrow 3:15 = 1:5$

2) $8:72 \rightarrow 8 \div 8 = 1, \ 72 \div 8 = 9 \rightarrow 8:72 = 1:9$

3) $15:25 \rightarrow 15 \div 5 = 3, \ 25 \div 5 = 5 \rightarrow 15:25 = 3:5$

4) $35:10 \rightarrow 35 \div 5 = 7, \ 10 \div 5 = 2 \rightarrow 35:10 = 7:2$

5) $36:42 \rightarrow 36 \div 6 = 6, \ 42 \div 6 = 7 \rightarrow 36:42 = 6:7$

6) $24:64 \rightarrow 24 \div 8 = 3, \ 64 \div 8 = 8 \rightarrow 24:64 = 3:8$

7) In Jack's class the ratio of tall students to short students is: $\frac{48}{20} = \frac{48 \div 4}{20 \div 4} = \frac{12}{5}$ and in Michael's class the ratio is: $\frac{28}{12} = \frac{28 \div 4}{12 \div 4} = \frac{7}{3}$. Compare two fractions: $\frac{12}{5} = \frac{12 \times 3}{5 \times 3} = \frac{36}{15}$ and $\frac{7}{3} = \frac{7 \times 5}{3 \times 5} = \frac{35}{15} \rightarrow \frac{36}{15} > \frac{35}{15}$. In Jack's class the ratio of tall to short students is higher.

8) Write a proportion and solve: $\frac{7}{7.49} = \frac{21}{x} \rightarrow x = 3 \times 7.49 = \$22.47$

9) $\frac{3}{4} = \frac{12}{x} \rightarrow 3 \times x = 4 \times 12 \rightarrow 3x = 48 \rightarrow x = \frac{48}{3} = 16$

10) $\frac{2}{5} = \frac{x}{20} \rightarrow 5 \times x = 2 \times 20 \rightarrow 5x = 40 \rightarrow x = \frac{40}{5} = 8$

11) $\frac{9}{x} = \frac{3}{5} \rightarrow 9 \times 5 = 3 \times x \rightarrow 45 = 3x \rightarrow x = \frac{45}{3} = 15$

12) $\frac{x}{24} = \frac{5}{6} \rightarrow 6 \times x = 5 \times 24 \rightarrow 6x = 120 \rightarrow x = \frac{120}{6} = 20$

13) $\frac{8}{4} = \frac{x}{16} \rightarrow 8 \times 16 = 4 \times x \rightarrow 128 = 4x \rightarrow x = \frac{128}{4} = 32$

14) $\frac{3}{x} = \frac{12}{16} \rightarrow 3 \times 16 = 12 \times x \rightarrow 48 = 12x \rightarrow x = \frac{48}{12} = 4$

15) $\frac{24}{32} = \frac{3}{x} \rightarrow 24 \times x = 3 \times 32 \rightarrow 24x = 96 \rightarrow x = \frac{96}{24} = 4$

16) $\frac{x}{7} = \frac{21}{49} \rightarrow 49 \times x = 7 \times 21 \rightarrow 49x = 147 \rightarrow x = \frac{147}{49} = 3$

17) $\frac{6}{10} = \frac{36}{x} \rightarrow 6 \times x = 36 \times 10 \rightarrow 6x = 360 \rightarrow x = \frac{360}{6} = 60 \rightarrow x = 60 \ feet$

18) $\frac{4.6}{7} = \frac{x}{28} \rightarrow 7 \times x = 28 \times 4.6 \rightarrow 7x = 128.8 \rightarrow x = \frac{128.8 \div 7}{7 \div 7} = 18.4 \rightarrow x = 18.4 \ meters$

19) Part = Percent × Base → 15% × 60 = $\frac{15}{100}$ × 60 = 0.15 × 60 = 9

20) Part = Percent × Base → 20% × 500 = $\frac{20}{100}$ × 500 = 0.2 × 500 = 100

21) Percent = Part ÷ Base → 25 ÷ 250 = $\frac{25}{250}$ = $\frac{25 \div 25}{250 \div 25}$ = $\frac{1}{10}$ = $\frac{1}{10}$ × 100 = 10%

22) Percent = Part ÷ Base → 30 ÷ 150 = $\frac{30}{150}$ = $\frac{30 \div 30}{150 \div 30}$ = $\frac{1}{5}$ = $\frac{1}{5}$ × 100 = 20%

23) Base = Part ÷ Percent → 15 ÷ 10% = 15 ÷ $\frac{10}{100}$ = 15 ÷ 0.1 = 150

24) Base = Part ÷ Percent → 25 ÷ 5% = 25 ÷ $\frac{5}{100}$ = 25 ÷ 0.05 = 500

25) Percent of change = $\frac{new\ number - original\ number}{original\ number}$ × 100 = $\frac{21-15}{15}$ × 100 = $\frac{6}{15}$ × 100 = 40%

26) $\frac{36-45}{45}$ × 100 = $\frac{-9}{45}$ × 100 = −20% (the negative sign means that the price decreased)

27) 5% × 500 = $\frac{5}{100}$ × 500 = 25 → $500 + $25 = $525

28) 14% × $500 = $\frac{14}{100}$ × $500 = $500 × 0.14 = $70

29) 10% × $60 = $\frac{10}{100}$ × $60 = $6 → $60 + $6 = $66

30) 800 × 3% × 2 = 800 × $\frac{3}{100}$ × 2 = $\frac{4,800}{100}$ = 48

31) 260 × 10% × 5 = 260 × $\frac{10}{100}$ × 5 = 260 × $\frac{1}{10}$ × 5 = 130

32) 380 × 4% × 3 = 380 × $\frac{4}{100}$ × 3 = 45.6

33) 1,200 × 2% × 1 = 1,200 × $\frac{2}{100}$ × 1 = 24

4 Exponents and Variables

Math topics that you'll learn in this chapter:

1. Multiplication Property of Exponents
2. Division Property of Exponents
3. Powers of Products and Quotients
4. Zero and Negative Exponents
5. Negative Exponents and Negative Bases
6. Scientific Notation
7. Radicals

41

Multiplication Property of Exponents

☆ Exponents are shorthand for repeated multiplication of the same number by itself. For example, instead of 2×2, we can write 2^2. For $3 \times 3 \times 3 \times 3$, we can write 3^4.

☆ In algebra, a variable is a letter used to stand for a number. The most common letters are: $x, y, z, a, b, c, m,$ and n.

☆ Exponent's rules: $x^a \times x^b = x^{a+b}$, $\dfrac{x^a}{x^b} = x^{a-b}$

$$(x^a)^b = x^{a \times b} \qquad (xy)^a = x^a \times y^a \qquad \left(\dfrac{a}{b}\right)^c = \dfrac{a^c}{b^c}$$

Examples:

Example 1. *Multiply.* $3x^2 \times 4x^3$

Solution: Use Exponent's rules: $x^a \times x^b = x^{a+b} \rightarrow x^2 \times x^3 = x^{2+3} = x^5$
Then: $3x^2 \times 4x^3 = 12x^5$

Example 2. *Simplify.* $(x^2 y^4)^3$

Solution: Use Exponent's rules: $(x^a)^b = x^{a \times b}$.
Then: $(x^2 y^4)^3 = x^{2 \times 3} y^{4 \times 3} = x^6 y^{12}$

Example 3. *Multiply.* $6x^7 \times 4x^3$

Solution: Use Exponent's rules: $x^a \times x^b = x^{a+b} \rightarrow x^7 \times x^3 = x^{7+3} = x^{10}$
Then: $6x^7 \times 4x^3 = 24x^{10}$

Example 4. *Simplify.* $\left(x^2 y^5\right)^4$

Solution: Use Exponent's rules: $(x^a)^b = x^{a \times b}$.
Then: $\left(x^2 y^5\right)^4 = x^{2 \times 4} y^{5 \times 4} = x^8 y^{20}$

Division Property of Exponents

For division of exponents use following formulas:

☆ $\frac{x^a}{x^b} = x^{a-b}$ $(x \neq 0)$

☆ $\frac{x^a}{x^b} = \frac{1}{x^{b-a}}$, $(x \neq 0)$

☆ $\frac{1}{x^b} = x^{-b}$

Examples:

Example 1. ***Simplify.*** $\frac{12x^2y}{3xy^3} =$

Solution: First, cancel the common factor: $3 \rightarrow \frac{12x^2y}{3xy^3} = \frac{4x^2y}{xy^3}$

Use Exponent's rules: $\frac{x^a}{x^b} = x^{a-b} \rightarrow \frac{x^2}{x} = x^{2-1} = x^1$ and $\frac{x^a}{x^b} = \frac{1}{x^{b-a}} \rightarrow \frac{y}{y^3} = \frac{1}{y^{3-1}} = \frac{1}{y^2}$

Then: $\frac{12x^2y}{3xy^3} = \frac{4x}{y^2}$

Example 2. ***Simplify.*** $\frac{48x^{12}}{16x^9} =$

Solution: Use Exponent's rules: $\frac{x^a}{x^b} = x^{a-b} \rightarrow \frac{x^{12}}{x^9} = x^{12-9} = x^3$

Then: $\frac{48x^{12}}{16x^9} = 3x^3$

Example 3. ***Simplify.*** $\frac{6x^5y^7}{42x^8y^2} =$

Solution: First, cancel the common factor: $6 \rightarrow \frac{x^5y^7}{7x^8y^2}$

Use Exponent's rules: $\frac{x^a}{x^b} = x^{a-b} \rightarrow \frac{x^5}{x^8} = x^{5-8} = x^{-3} = \frac{1}{x^3}$ and $\frac{y^7}{y^2} = y^{7-2} = y^5$

Then: $\frac{6x^5y^7}{42x^8y^2} = \frac{y^5}{7x^3}$

Powers of Products and Quotients

☆ For any nonzero numbers a and b and any integer x, $(ab)^x = a^x \times b^x$

and $\left(\dfrac{a}{b}\right)^c = \dfrac{a^c}{b^c}$

Examples:

Example 1. Simplify. $\left(6x^4y^5\right)^2$

Solution: Use Exponent's rules: $(x^a)^b = x^{a \times b}$

$\left(6x^4y^5\right)^2 = (6)^2(x^4)^2(y^5)^2 = 36x^{4 \times 2}y^{5 \times 2} = 36x^8y^{10}$

Example 2. Simplify. $\left(\dfrac{5x^5}{2x^4}\right)^2$

Solution: First, cancel the common factor: $x^4 \rightarrow \left(\dfrac{5x^5}{2x^4}\right) = \left(\dfrac{5x}{2}\right)^2$

Use Exponent's rules: $\left(\dfrac{a}{b}\right)^c = \dfrac{a^c}{b^c}$. Then: $\left(\dfrac{5x}{2}\right)^2 = \dfrac{(5x)^2}{(2)^2} = \dfrac{25x^2}{4}$

Example 3. Simplify. $(-6x^7y^3)^2$

Solution: Use Exponent's rules: $(x^a)^b = x^{a \times b}$

$(-6x^7y^3)^2 = (-6)^2(x^7)^2(y^3)^2 = 36x^{7 \times 2}y^{3 \times 2} = 36x^{14}y^6$

Example 4. Simplify. $\left(\dfrac{8x^3}{5x^7}\right)^2$

Solution: First, cancel the common factor: $x^3 \rightarrow \left(\dfrac{8x^3}{5x^7}\right)^2 = \left(\dfrac{8}{5x^4}\right)^2$

Use Exponent's rules: $\left(\dfrac{a}{b}\right)^c = \dfrac{a^c}{b^c}$, Then: $\left(\dfrac{8}{5x^4}\right)^2 = \dfrac{8^2}{\left(5x^4\right)^2} = \dfrac{64}{25x^8}$

Zero and Negative Exponents

☆ Zero-Exponent Rule: $a^0 = 1$, this means that anything raised to the zero power is 1. For example: $(5xy)^0 = 1$ (number zero is an exception: $\mathbf{0^0 = 0}$)

☆ A negative exponent simply means that the base is on the wrong side of the fraction line, so you need to flip the base to the other side. For instance, "x^{-2}" (pronounced as "ecks to the minus two") just means "x^2" but underneath, as in $\frac{1}{x^2}$.

Examples:

Example 1. *Evaluate.* $\left(\frac{3}{7}\right)^{-2} =$

Solution: Use negative exponent's rule: $\left(\frac{x^a}{x^b}\right)^{-2} = \left(\frac{x^b}{x^a}\right)^2 \rightarrow \left(\frac{3}{7}\right)^{-2} = \left(\frac{7}{3}\right)^2$
Then: $\left(\frac{7}{3}\right)^2 = \frac{7^2}{3^2} = \frac{49}{9}$

Example 2. *Evaluate.* $\left(\frac{2}{5}\right)^{-3} =$

Solution: Use negative exponent's rule: $\left(\frac{x^a}{x^b}\right)^{-3} = \left(\frac{x^b}{x^a}\right)^3 \rightarrow \left(\frac{2}{5}\right)^{-3} = \left(\frac{5}{2}\right)^3 =$
Then: $\left(\frac{5}{2}\right)^3 = \frac{5^3}{2^3} = \frac{125}{8}$

Example 3. *Evaluate.* $\left(\frac{a}{b}\right)^0 =$

Solution: Use zero-exponent Rule: $a^0 = 1$
Then: $\left(\frac{a}{b}\right)^0 = 1$

Example 4. *Evaluate.* $\left(\frac{9}{5}\right)^{-1} =$

Solution: Use negative exponent's rule: $\left(\frac{x^a}{x^b}\right)^{-1} = \left(\frac{x^b}{x^a}\right)^1 \rightarrow \left(\frac{9}{5}\right)^{-1} = \left(\frac{5}{9}\right)^1 = \frac{5}{9}$

Negative Exponents and Negative Bases

☆ A negative exponent is the reciprocal of that number with a positive exponent. $(3)^{-2} = \frac{1}{3^2}$

☆ To simplify a negative exponent, make the power positive!

☆ The parenthesis is important! -5^{-2} is not the same as $(-5)^{-2}$

$$-5^{-2} = -\frac{1}{5^2} \text{ and } (-5)^{-2} = +\frac{1}{5^2}$$

Examples:

Example 1. Simplify. $\left(\frac{4a}{7c}\right)^{-2} =$

Solution: Use negative exponent's rule: $\left(\frac{x^a}{x^b}\right)^{-2} = \left(\frac{x^b}{x^a}\right)^2 \rightarrow \left(\frac{4a}{7c}\right)^{-2} = \left(\frac{7c}{4a}\right)^2$
Now use exponent's rule: $\left(\frac{a}{b}\right)^c = \frac{a^c}{b^c} \rightarrow \left(\frac{7c}{4a}\right)^2 = \frac{7^2c^2}{4^2a^2}$
Then: $\frac{7^2c^2}{4^2a^2} = \frac{49c^2}{16a^2}$

Example 2. Simplify. $\left(\frac{3x}{y}\right)^{-3} =$

Solution: Use negative exponent's rule: $\left(\frac{x^a}{x^b}\right)^{-3} = \left(\frac{x^b}{x^a}\right)^3 \rightarrow \left(\frac{3x}{y}\right)^{-3} = \left(\frac{y}{3x}\right)^3$
Now use exponent's rule: $\left(\frac{a}{b}\right)^c = \frac{a^c}{b^c} \rightarrow \left(\frac{y}{3x}\right)^3 = \frac{y^3}{3^3x^3} = \frac{y^3}{27x^3}$

Example 3. Simplify. $\left(\frac{7a}{4c}\right)^{-2} =$

Solution: Use negative exponent's rule: $\left(\frac{x^a}{x^b}\right)^{-2} = \left(\frac{x^b}{x^a}\right)^2 \rightarrow \left(\frac{7a}{4c}\right)^{-2} = \left(\frac{4c}{7a}\right)^2$
Now use exponent's rule: $\left(\frac{a}{b}\right)^c = \frac{a^c}{b^c} \rightarrow \left(\frac{4c}{7a}\right)^2 = \frac{4^2c^2}{7^2a^2}$
Then: $\frac{4^2c^2}{7^2a^2} = \frac{16c^2}{49a^2}$

Scientific Notation

★ Scientific notation is used to write very big or very small numbers in decimal form.

★ In scientific notation, all numbers are written in the form of: $m \times 10^n$, where m is greater than 1 and less than 10.

★ To convert a number from scientific notation to standard form, move the decimal point to the left (if the exponent of ten is a negative number), or to the right (if the exponent is positive).

Examples:

Example 1. Write 0.000037 **in scientific notation.**

Solution: First, move the decimal point to the right so you have a number between 1 and 10. That number is 3.7. Now, determine how many places the decimal moved in step 1 by the power of 10. We moved the decimal point 5 digits to the right. Then: 10^{-5}. When the decimal moved to the right, the exponent is negative. Then: $0.000037 = 3.7 \times 10^{-5}$

Example 2. Write 5.3×10^{-3} **in standard notation.**

Solution: The exponent is negative 3. Then, move the decimal point to the left three digits. (remember $5.3 = 0000005.3$) When the decimal moved to the right, the exponent is negative. Then: $5.3 \times 10^{-3} = 0.0053$

Example 3. Write 0.00042 **in scientific notation.**

Solution: First, move the decimal point to the right so you have a number between 1 and 10. Then: $m = 4.2$. Now, determine how many places the decimal moved in step 1 by the power of 10. 10^{-4}. Then: $0.00042 = 4.2 \times 10^{-4}$

Example 4. Write 7.3×10^7 **in standard notation.**

Solution: The exponent is positive 7. Then, move the decimal point to the right seven digits. (remember $7.3 = 7.3000000$) Then: $7.3 \times 10^7 = 73,000,000$

Radicals

★ If n is a positive integer and x is a real number, then: $\sqrt[n]{x} = x^{\frac{1}{n}}$,

$$\sqrt[n]{xy} = x^{\frac{1}{n}} \times y^{\frac{1}{n}}, \sqrt[n]{\frac{x}{y}} = \frac{x^{\frac{1}{n}}}{y^{\frac{1}{n}}}, \text{ and } \sqrt[n]{x} \times \sqrt[n]{y} = \sqrt[n]{xy}$$

★ A square root of x is a number r whose square is: $r^2 = x$ (r is a square root of x)

★ To add and subtract radicals, we need to have the same values under the radical. For example: $\sqrt{5} + 3\sqrt{5} = 4\sqrt{5}$, $5\sqrt{6} - 2\sqrt{6} = 3\sqrt{5}$

Examples:

Example 1. Find the square root of $\sqrt{256}$.

Solution: First, factor the number: $256 = 16^2$. Then: $\sqrt{256} = \sqrt{16^2}$
Now use radical rule: $\sqrt[n]{a^n} = a$. Then: $\sqrt{256} = \sqrt{16^2} = 16$

Example 2. Evaluate. $\sqrt{9} \times \sqrt{36} =$

Solution: Find the values of $\sqrt{9}$ and $\sqrt{36}$. Then: $\sqrt{9} \times \sqrt{36} = 3 \times 6 = 18$

Example 3. Solve. $3\sqrt{7} + 11\sqrt{7}$.

Solution: Since we have the same values under the radical, we can add these two radicals: $3\sqrt{7} + 11\sqrt{7} = 14\sqrt{7}$

Example 4. Evaluate. $\sqrt{40} \times \sqrt{10} =$

Solution: Use this radical rule: $\sqrt[n]{x} \times \sqrt[n]{y} = \sqrt[n]{xy} \rightarrow \sqrt{40} \times \sqrt{10} = \sqrt{400}$
The square root of 400 is 20. Then: $\sqrt{40} \times \sqrt{10} = \sqrt{400} = 20$

Day 4: Practices

✍ Find the products.

1) $5x^3 \times 2x =$

2) $x^4 \times 5x^2y =$

3) $2xy \times 3x^5y^2 =$

4) $4xy^2 \times 2x^2y =$

5) $-3x^3y^3 \times 2x^2y^2 =$

6) $-5xy^2 \times 3x^5y^2 =$

7) $-5x^2y^6 \times 6x^5y^2 =$

8) $-2x^3y^3 \times 2x^3y^3 =$

9) $-7xy^3 \times 4x^5y^2 =$

10) $-x^4y^3 \times (-5x^6y^2) =$

11) $-6y^6 \times 7x^6y^2 =$

12) $-8x^4 \times 2y^2 =$

✍ Simplify.

13) $\frac{3^2 \times 3^3}{3^3 \times 3} =$

14) $\frac{4^2 \times 4^4}{5^4 \times 5} =$

15) $\frac{14x^5}{7x^2} =$

16) $\frac{15x^3}{5x^6} =$

17) $\frac{64y^3}{8xy^7} =$

18) $\frac{10x^4y^5}{30x^5y^4} =$

19) $\frac{11y}{44x^3y^3} =$

20) $\frac{40xy^3}{120xy^3} =$

21) $\frac{45x^3}{25xy^3} =$

22) $\frac{72y^6x}{36x^8y^9} =$

✍ Solve.

23) $(x^2 \, y^2)^3 =$

24) $(2x^3 \, y^2)^3 =$

25) $(2x \times 3xy^2)^2 =$

26) $(4x \times 2y^4)^2 =$

27) $\left(\frac{3x}{x^2}\right)^2 =$

28) $\left(\frac{6y}{18y^3}\right)^2 =$

29) $\left(\frac{3x^2y^2}{12x^4y^3}\right)^3 =$

30) $\left(\frac{23x^5y^3}{46x^3y^5}\right)^3 =$

31) $\left(\frac{16x^7y^3}{48x^5y^2}\right)^2 =$

32) $\left(\frac{12x^5y^6}{60x^7y^2}\right)^2 =$

Evaluate each expression. (Zero and Negative Exponents)

33) $\left(\frac{1}{3}\right)^{-2} =$

34) $\left(\frac{1}{4}\right)^{-3} =$

35) $\left(\frac{1}{6}\right)^{-2} =$

36) $\left(\frac{2}{3}\right)^{-3} =$

37) $\left(\frac{2}{5}\right)^{-3} =$

38) $\left(\frac{3}{5}\right)^{-2} =$

Write each expression with positive exponents.

39) $2y^{-3} =$

40) $13y^{-5} =$

41) $-20x^{-2} =$

42) $15a^{-2}b^3 =$

43) $23a^2b^{-4}c^{-8} =$

44) $-4x^4y^{-2} =$

45) $\frac{16y}{x^3y^{-4}} =$

46) $\frac{30a^{-3}b}{-100c^{-2}} =$

Write each number in scientific notation.

47) $0.00518 =$

48) $0.000042 =$

49) $78,000 =$

50) $92,000,000 =$

Evaluate.

51) $\sqrt{5} \times \sqrt{5} =$

52) $\sqrt{25} - \sqrt{4} =$

53) $\sqrt{81} + \sqrt{36} =$

54) $\sqrt{4} \times \sqrt{25} =$

55) $\sqrt{2} \times \sqrt{18} =$

56) $4\sqrt{2} + 3\sqrt{2} =$

57) $5\sqrt{7} + 2\sqrt{7} =$

58) $\sqrt{45} + 2\sqrt{5} =$

Day 4: Answers

1) $5x^3 \times 2x \to x^3 \times x^1 = x^{3+1} = x^4 \to 5x^3 \times 2x = 10x^4$

2) $x^4 \times 5x^2y \to x^4 \times x^2 = x^{4+2} = x^6 \to x^4 \times 5x^2y = 5x^6y$

3) $2xy \times 3x^5y^2 \to x \times x^5 = x^{1+5} = x^6,\ y \times y^2 = y^{1+2} = y^3 \to 2xy \times 3x^5y^2 = 6x^6y^3$

4) $4xy^2 \times 2x^2y \to x \times x^2 = x^{1+2} = x^3,\ y^2 \times y = y^{2+1} = y^3 \to 4xy^2 \times 2x^2y = 8x^3y^3$

5) $-3x^3y^3 \times 2x^2y^2 \to x^3 \times x^2 = x^{3+2} = x^5,\ y^3 \times y^2 = y^{3+2} = y^5 \to$
   $-3x^3y^3 \times 2x^2y^2 = -6x^5y^5$

6) $-5xy^2 \times 3x^5y^2 \to x \times x^5 = x^{1+5} = x^6,\ y^2 \times y^2 = y^{2+2} = y^4 \to$
   $-5xy^2 \times 3x^5y^2 = -15x^6y^4$

7) $-5x^2y^6 \times 6x^5y^2 \to x^2 \times x^5 = x^{2+5} = x^7,\ y^6 \times y^2 = y^{6+2} = y^8 \to$
   $-5x^2y^6 \times 6x^5y^2 = -30x^7y^8$

8) $-2x^3y^3 \times 2x^3y^3 \to x^3 \times x^3 = x^{3+3} = x^6,\ y^3 \times y^3 = y^{3+3} = y^6 \to$
   $-2x^3y^3 \times 2x^3y^3 = -4x^6y^6$

9) $-7xy^3 \times 4x^5y^2 \to x \times x^5 = x^{1+5} = x^6,\ y^3 \times y^2 = y^{3+2} = y^5 \to$
   $-7xy^3 \times 4x^5y^2 = -28x^6y^5$

10) $-x^4y^3 \times (-5x^6y^2) \to x^4 \times x^6 = x^{4+6} = x^{10},\ y^3 \times y^2 = y^{3+2} = y^5 \to$
   $-x^4y^3 \times (-5x^6y^2) = 5x^{10}y^5$

11) $-6y^6 \times 7x^6y^2 \to y^6 \times y^2 = y^{6+2} = y^8 \to -6y^6 \times 7x^6y^2 = -42x^6y^8$

12) $-8x^4 \times 2y^2 = -16x^4y^2$

13) $\dfrac{3^2 \times 3^3}{3^3 \times 3} = \dfrac{3^{2+3}}{3^{3+1}} = \dfrac{3^5}{3^4} = 3^{5-4} = 3^1 = 3$

14) $\dfrac{4^2 \times 4^4}{5^4 \times 5} = \dfrac{4^{2+4}}{5^{4+1}} = \dfrac{4^6}{5^5}$

15) $\dfrac{14x^5}{7x^2} \to \dfrac{14 \div 7}{7 \div 7} = 2,\ \dfrac{x^5}{x^2} = x^{5-2} = x^3 \to \dfrac{14x^5}{7x^2} = 2x^3$

16) $\dfrac{15x^3}{5x^6} \to \dfrac{15 \div 5}{5 \div 5} = 3,\ \dfrac{x^3}{x^6} = x^{3-6} = x^{-3} = \dfrac{1}{x^3} \to \dfrac{15x^3}{5x^6} = \dfrac{3}{x^3}$

17) $\dfrac{64y^3}{8xy^7} \to \dfrac{64 \div 8}{8 \div 8} = 8,\ \dfrac{y^3}{y^7} = y^{3-7} = y^{-4} = \dfrac{1}{y^4} \to \dfrac{64y^3}{8xy^7} = \dfrac{8}{xy^4}$

18) $\dfrac{10x^4y^5}{30x^5y^4} \to \dfrac{10 \div 10}{30 \div 10} = \dfrac{1}{3},\ \dfrac{x^4}{x^5} = x^{4-5} = x^{-1} = \dfrac{1}{x},\ \dfrac{y^5}{y^4} = y^{5-4} = y^1 \to \dfrac{10x^4y^5}{30x^5y^4} = \dfrac{y}{3x}$

19) $\frac{11y}{44x^3y^3} \rightarrow \frac{11\div 11}{44\div 11} = \frac{1}{4}, \frac{y}{y^3} = y^{1-3} = y^{-2} = \frac{1}{y^2} \rightarrow \frac{11y}{44x^3y^3} = \frac{1}{4x^3y^2}$

20) $\frac{40xy^3}{120xy^3} \rightarrow \frac{40\div 40}{120\div 40} = \frac{1}{3}, \frac{x}{x} = x^{1-1} = x^0 = 1, \frac{y^3}{y^3} = y^{3-3} = y^0 = 1 \rightarrow \frac{40xy^3}{120xy^3} = \frac{1}{3}$

21) $\frac{45x^3}{25xy^3} \rightarrow \frac{45\div 5}{25\div 5} = \frac{9}{5}, \frac{x^3}{x} = x^{3-1} = x^2 \rightarrow \frac{45x^3}{25xy^3} = \frac{9x^2}{5y^3}$

22) $\frac{72y^6x}{36x^8y^9} \rightarrow \frac{72\div 36}{36\div 36} = \frac{2}{1} = 2, \frac{y^6}{y^9} = y^{6-9} = y^{-3} = \frac{1}{y^3}, \frac{x}{x^8} = x^{1-8} = x^{-7} = \frac{1}{x^7} \rightarrow \frac{72y^6x}{36x^8y^9} = \frac{2}{x^7y^3}$

23) $(x^2\,y^2)^3 = x^{2\times 3}\,y^{2\times 3} = x^6\,y^6$

24) $(2x^3\,y^2)^3 = 2^3 x^{3\times 3}\,y^{2\times 3} = 8x^9y^6$

25) $(2x \times 3xy^2)^2 = (2\times 3)^2 x^{(1+1)\times 2}\,y^{2\times 2} = 6^2 x^{2\times 2}\,y^4 = 36x^4\,y^4$

26) $(4x \times 2y^4)^2 = (4\times 2)^2 x^{1\times 2}\,y^{4\times 2} = 8^2 x^2\,y^8 = 64x^2\,y^8$

27) $\left(\frac{3x}{x^2}\right)^2 = \frac{3^{1\times 2}x^{1\times 2}}{x^{2\times 2}} = \frac{3^2 x^2}{x^4} \rightarrow 3^2 = 9, \frac{x^2}{x^4} = x^{2-4} = x^{-2} = \frac{1}{x^2} \rightarrow \left(\frac{3x}{x^2}\right)^2 = \frac{9}{x^2}$

28) $\left(\frac{6y}{18y^3}\right)^2 = \frac{6^{1\times 2}y^{1\times 2}}{(18)^{1\times 2}y^{3\times 2}} = \frac{6^2 y^2}{18^2 y^6} \rightarrow \frac{36}{324} = \frac{1}{9}, \frac{y^2}{y^6} = y^{2-6} = y^{-4} = \frac{1}{y^4} \rightarrow \left(\frac{6y}{18y^3}\right)^2 = \frac{1}{9y^4}$

29) $\left(\frac{3x^2y^2}{12x^4y^3}\right)^3 = \frac{3^{1\times 3}x^{2\times 3}y^{2\times 3}}{(12)^{1\times 3}x^{4\times 3}y^{3\times 3}} = \frac{3^3 x^6 y^6}{12^3 x^{12} y^9} \rightarrow \frac{3^3}{12^3} = \frac{27}{1,728} = \frac{1}{64}, \frac{x^6}{x^{12}} = x^{6-12} = x^{-6} = \frac{1}{x^6}, \frac{y^6}{y^9} =$

$y^{6-9} = y^{-3} = \frac{1}{y^3} \rightarrow \left(\frac{3x^2y^2}{12x^4y^3}\right)^3 = \frac{1}{64x^6y^3}$

30) $\left(\frac{23x^5y^3}{46x^3y^5}\right)^3 = \frac{23^{1\times 3}x^{5\times 3}y^{3\times 3}}{(23\times 2)^{1\times 3}x^{3\times 3}y^{5\times 3}} = \frac{23^3 x^{15} y^9}{23^3\times 2^3 x^9 y^{15}} = \frac{x^{15} y^9}{8x^9 y^{15}} \rightarrow \frac{x^{15}}{x^9} = x^{15-9} = x^6, \frac{y^9}{y^{15}} = y^{9-15} =$

$y^{-6} = \frac{1}{y^6} \rightarrow \left(\frac{23x^5y^3}{46x^3y^5}\right)^3 = \frac{x^6}{8y^6}$

31) $\left(\frac{16x^7y^3}{48x^5y^2}\right)^2 = \frac{16^{1\times 2}x^{7\times 2}y^{3\times 2}}{(16\times 3)^{1\times 2}x^{5\times 2}y^{2\times 2}} = \frac{16^2 x^{14} y^6}{16^2\times 3^2 x^{10} y^4} = \frac{x^{14} y^6}{9x^{10} y^4} \rightarrow \frac{x^{14}}{x^{10}} = x^{14-10} = x^4, \frac{y^6}{y^4} = y^{6-4} =$

$y^2 \rightarrow \left(\frac{16x^7y^3}{48x^5y^2}\right)^2 = \frac{x^4y^2}{9}$

32) $\left(\frac{12x^5y^6}{60x^7y^2}\right)^2 = \frac{12^{1\times 2}x^{5\times 2}y^{6\times 2}}{(12\times 5)^{1\times 2}x^{7\times 2}y^{2\times 2}} = \frac{12^2 x^{10} y^{12}}{12^2\times 5^2 x^{14} y^4} = \frac{x^{10} y^{12}}{25x^{14} y^4} \rightarrow \frac{x^{10}}{x^{14}} = x^{10-14} = x^{-4} = \frac{1}{x^4}, \frac{y^{12}}{y^4} =$

$y^{12-4} = y^8 \rightarrow \left(\frac{12x^5y^6}{60x^7y^2}\right)^2 = \frac{y^8}{25x^4}$

33) $\left(\frac{1}{3}\right)^{-2} = 3^2 = 9$ 34) $\left(\frac{1}{4}\right)^{-3} = 4^3 = 64$

- *www.EffortlessMath.com*

35) $\left(\frac{1}{6}\right)^{-2} = 6^2 = 36$

36) $\left(\frac{2}{3}\right)^{-3} = \left(\frac{3}{2}\right)^3 = \frac{3^3}{2^3} = \frac{27}{8}$

37) $\left(\frac{2}{5}\right)^{-3} = \left(\frac{5}{2}\right)^3 = \frac{5^3}{2^3} = \frac{125}{8}$

38) $\left(\frac{3}{5}\right)^{-2} = \left(\frac{5}{3}\right)^2 = \frac{5^2}{3^2} = \frac{25}{9}$

39) $2y^{-3} = \frac{2}{y^3}$

40) $13y^{-5} = \frac{13}{y^5}$

41) $-20x^{-2} = -\frac{20}{x^2}$

42) $15a^{-2}b^3 = \frac{15b^3}{a^2}$

43) $23a^2b^{-4}c^{-8} = \frac{23a^2}{b^4c^8}$

44) $-4x^4y^{-2}2^{-7} = -\frac{4x^4}{y^2}$

45) $\frac{16y}{x^3y^{-4}} \rightarrow \frac{y}{y^{-4}} = y^{1-(-4)} = y^5 \rightarrow \frac{16y}{x^3y^{-4}} = \frac{16y^5}{x^3}$

46) $\frac{30a^{-3}b}{-100c^{-2}} \rightarrow -\frac{30 \div 10}{100 \div 10} = -\frac{3}{10} \rightarrow \frac{30a^{-3}b}{-100c^{-2}} = -\frac{3c^2b}{10a^3}$

47) $0.00518 = 5.18 \times 10^{-3}$

48) $0.000042 = 4.2 \times 10^{-5}$

49) $78,000 = 7.8 \times 10^4$

50) $92,000,000 = 9.2 \times 10^7$

51) $\sqrt{5} \times \sqrt{5} = 5$

52) $\sqrt{25} - \sqrt{4} = 5 - 2 = 3$

53) $\sqrt{81} + \sqrt{36} = 9 + 6 = 15$

54) $\sqrt{4} \times \sqrt{25} = 2 + 5 = 10$

55) $\sqrt{2} \times \sqrt{18} = \sqrt{36} = 6$

56) $4\sqrt{2} + 3\sqrt{2} = 7\sqrt{2}$

57) $5\sqrt{7} + 2\sqrt{7} = 7\sqrt{7}$

58) $\sqrt{45} + 2\sqrt{5} = \sqrt{5 \times 9} + 2\sqrt{5} = 3\sqrt{5} + 2\sqrt{5} = 5\sqrt{5}$

5 Expressions and Variables

Math topics that you'll learn in this chapter:

1. Simplifying Variable Expressions
2. Simplifying Polynomial Expressions
3. The Distributive Property
4. Evaluating One Variable
5. Evaluating Two Variables

55

Simplifying Variable Expressions

☆ In algebra, a variable is a letter used to stand for a number. The most common letters are $x, y, z, a, b, c, m, and\ n$.

☆ An algebraic expression is an expression that contains integers, variables, and math operations such as addition, subtraction, multiplication, division, etc.

☆ In an expression, we can combine "like" terms. (values with same variable and same power)

Examples:

Example 1. Simplify. $(3x + 9x + 2) =$

Solution: In this expression, there are three terms: $3x,\ 9x$, and 2. Two terms are "like terms": $3x$ and $9x$. Combine like terms. $3x + 9x = 12x$. Then: $(3x + 9x + 2) = 12x + 2$ (***remember you cannot combine variables and numbers.***)

Example 2. Simplify. $-17x^2 + 6x + 15x^2 - 13 =$

Solution: Combine "like" terms: $-17x^2 + 15x^2 = -2x^2$
Then: $-17x^2 + 6x + 15x^2 - 13 = -2x^2 + 6x - 13$

Example 3. Simplify. $5x - 18 - 6x^2 + 3x^2 =$

Solution: Combine like terms. Then:
$5x - 18 - 6x^2 + 3x^2 = -3x^2 + 5x - 18$

Example 4. Simplify. $-5x - 4x^2 + 9x - 11x^2 =$

Solution: Combine "like" terms: $-5x + 9x = 4x$, and $-4x^2 - 11x^2 = -15x^2$
Then: $-5x - 4x^2 + 9x - 11x^2 = 4x - 15x^2$. Write in standard form (biggest powers first): $4x - 15x^2 = -15x^2 + 4x$

Simplifying Polynomial Expressions

☆ In mathematics, a polynomial is an expression consisting of variables and coefficients that involves only the operations of addition, subtraction, multiplication, and non–negative integer exponents of variables.

$$P(x) = a_n x^n + a_{n-1} x^{n-1} + \dots + a_2 x^2 + a_1 x + a_0$$

☆ Polynomials must always be simplified as much as possible. It means you must add together any like terms. (values with same variable and same power)

Examples:

Example 1. Simplify this Polynomial Expressions. $-2x^2 + 9x^3 + 5x^3 - 7x^4$

Solution: Combine "like" terms: $9x^3 + 5x^3 = 14x^3$

Then: $-2x^2 + 9x^3 + 5x^3 - 7x^4 = -2x^2 + 14x^3 - 7x^4$
Now, write the expression in standard form:
$$-2x^2 + 14x^3 - 7x^4 = -7x^4 + 14x^3 - 2x^2$$

Example 2. Simplify this expression. $(4x^2 - x^3) - (-6x^3 + 3x^2) =$

Solution: First, multiply $(-)$ into $(-6x^3 + 3x^2)$:

$(4x^2 - x^3) - (-6x^3 + 3x^2) = 4x^2 - x^3 + 6x^3 - 3x^2$
Then combine "like" terms: $4x^2 - x^3 + 6x^3 - 3x^2 = x^2 + 5x^3$
And write in standard form: $x^2 + 5x^3 = 5x^3 + x^2$

Example 3. Simplify. $-2x^3 + 6x^4 - 5x^2 - 14x^4 =$

Solution: Combine "like" terms: $6x^4 - 14x^4 = -8x^4$
Then: $-2x^3 + 6x^4 - 5x^2 - 14x^4 = -2x^3 - 8x^4 - 5x^2$
And write in standard form: $-2x^3 - 8x^4 - 5x^2 = -8x^4 - 2x^3 - 5x^2$

The Distributive Property

☆ The distributive property (or the distributive property of multiplication over addition and subtraction) simplifies and solves expressions in the form of: $a(b + c)$ or $a(b - c)$

☆ The distributive property is multiplying a term outside the parentheses by the terms inside.

☆ Distributive Property rule: $a(b + c) = ab + ac$

Examples:

Example 1. Simply using the distributive property. $(3)(4x - 9)$

Solution: Use Distributive Property rule: $a(b + c) = ab + ac$

$(3)(4x - 9) = (3 \times 4x) + (3) \times (-9) = 12x - 27$

Example 2. Simply. $(-4)(-3x + 8)$

Solution: Use Distributive Property rule: $a(b + c) = ab + ac$

$(-4)(-3x + 8) = (-4 \times (-3x)) + (-4) \times (8) = 12x - 32$

Example 3. Simply. $(5)(3x + 4) - 13x$

Solution: First, simplify $(5)(3x + 4)$ using the distributive property.

Then: $(5)(3x + 4) = 15x + 20$

Now combine like terms: $(5)(3x + 4) - 13x = 15x + 20 - 13x$

In this expression, $15x$ and $-13x$ are "like terms" and we can combine them.

$15x - 13x = 2x$. Then: $15x + 20 - 13x = 2x + 20$

Evaluating One Variable

☆ To evaluate one variable expressions, find the variable and substitute a number for that variable.

☆ Perform the arithmetic operations.

Examples:

Example 1. Calculate this expression for $x = 1$. $\quad 9 + 8x$

Solution: First, substitute 1 for x.
Then: $9 + 8x = 9 + 8(1)$
Now, use order of operation to find the answer: $9 + 8(1) = 9 + 8 = 17$

Example 2. Evaluate this expression for $x = -2$. $\quad 7x - 3$

Solution: First, substitute -2 for x.
Then: $7x - 3 = 7(-2) - 3$
Now, use order of operation to find the answer: $7(-2) - 3 = -14 - 3 = -17$

Example 3. Find the value of this expression when $x = 3$. $(12 - 2x)$

Solution: First, substitute 3 for x,
Then: $12 - 2x = 12 - 2(3) = 12 - 6 = 6$

Example 4. Solve this expression for $x = -4$. $\quad 11 + 5x$

Solution: Substitute -4 for x.
Then: $11 + 5x = 11 + 5(-4) = 11 - 20 = -9$

Evaluating Two Variables

☆ To evaluate an algebraic expression, substitute a number for each variable.

☆ Perform the arithmetic operations to find the value of the expression.

Examples:

Example 1. Calculate this expression for $a = -2$ and $b = 3$. $(2a - 6b)$

Solution: First, substitute -2 for a, and 3 for b.
Then: $2a - 6b = 2(-2) - 6(3)$
Now, use order of operation to find the answer: $2(-2) - 6(3) = -4 - 18 = -22$

Example 2. Evaluate this expression for $x = -3$ and $y = 4$. $(2x - 4y)$

Solution: Substitute -3 for x, and 4 for y.
Then: $2x - 4y = 2(-3) - 4(4) = -6 - 16 = -22$

Example 3. Find the value of this expression $3(-4a + 2b)$, when $a = -2$ and $b = -3$.

Solution: Substitute -2 for a, and -3 for b.
Then: $3(-4a + 2b) = 3\big(-4(-2) + 2(-3)\big) = 3(8 - 6) = 3(2) = 6$

Example 4. Evaluate this expression. $-5x - 3y$, $x = 2$, $y = -6$

Solution: Substitute 2 for x, and -6 for y and simplify.
Then: $-5x - 3y = -5(2) - 3(-6) = -10 + 18 = 8$

Day 5: Practices

✍ Simplify each expression.

1) $2 - 3x - 1 =$

2) $-6 - 2x + 8 =$

3) $11x - 6x - 4 =$

4) $-16x + 25x - 5 =$

5) $5x + 5 - 15x =$

6) $4 + 5x - 6x - 5 =$

7) $3x + 10 - 2x - 20 =$

8) $-3 - 2x^2 - 5 + 3x =$

9) $-7 + 9x^2 - 2 + 2x =$

10) $4x^2 + 2x - 12x - 5 =$

11) $2x^2 - 3x - 5x + 6 - 9 =$

12) $x^2 - 6x - x + 2 - 3 =$

13) $10x^2 - x - 8x + 3 - 10 =$

14) $4x^2 - 7x - x^2 + 2x + 5 =$

✍ Simplify each polynomial.

15) $4x^2 + 3x^3 - x^2 + x =$

16) $5x^4 + x^5 - x^4 + 4x^2 =$

17) $15x^3 + 12x - 6x^2 - 9x^3 =$

18) $(7x^3 - 2x^2) + (6x^2 - 13x) =$

19) $(9x^4 + 6x^3) + (11x^3 - 5x^4) =$

20) $(15x^5 - 5x^3) - (4x^3 + 6x^2) =$

21) $(15x^4 + 7x^3) - (3x^3 - 26) =$

22) $(22x^4 + 6x^3) - (-2x^3 - 4x^4) =$

23) $(x^2 + 6x^3) + (-19x^2 + 6x^3) =$

24) $(2x^4 - x^3) + (-5x^3 - 7x^4) =$

✒️ **Use the distributive property to simply each expression.**

25) $3(5 + x) =$

26) $5(4 - x) =$

27) $6(2 - 5x) =$

28) $(4 - 3x)7 =$

29) $8(3 - 3x) =$

30) $(-1)(-6 + 2x) =$

31) $(-5)(3x - 3) =$

32) $(-x + 10)(-3)$

33) $(-2)(2 - 6x) =$

34) $(-6x - 4)(-7) =$

✒️ **Evaluate each expression using the value given.**

35) $x = 3 \rightarrow 12 - x =$

36) $x = 5 \rightarrow x + 7 =$

37) $x = 3 \rightarrow 3x - 5 =$

38) $x = 2 \rightarrow 18 - 3x =$

39) $x = 7 \rightarrow 5x - 4 =$

40) $x = 6 \rightarrow 21 - x =$

41) $x = 5 \rightarrow 10x - 20 =$

42) $x = -5 \rightarrow 4 - x =$

43) $x = -2 \rightarrow 25 - 3x =$

44) $x = -7 \rightarrow 16 - x =$

45) $x = -13 \rightarrow 40 - 2x =$

46) $x = -4 \rightarrow 20x - 6 =$

47) $x = -6 \rightarrow -11x - 19 =$

48) $x = -8 \rightarrow -1 - 3x =$

✒️ **Evaluate each expression using the values given.**

49) $x = 3, \ y = 2 \rightarrow 3x + 2y =$

50) $a = 4, \ b = 1 \rightarrow 2a - 6b =$

51) $x = 5, \ y = 7 \rightarrow 2x - 4y - 5 =$

52) $a = -3, \ b = 4 \rightarrow -3a + 4b + 2 =$

53) $x = -4, \ y = -3 \rightarrow 2x - 6 - 4y =$

Day 5: Answers

1) $2 - 3x - 1 \rightarrow 2 - 1 = 1 \rightarrow 2 - 3x - 1 = -3x + 1$

2) $-6 - 2x + 8 \rightarrow -6 + 8 = 2 \rightarrow -6 - 2x + 8 = -2x + 2$

3) $11x - 6x - 4 \rightarrow 11x - 6x = 5x \rightarrow 11x - 6x - 4 = 5x - 4$

4) $-16x + 25x - 5 \rightarrow -16x + 25x = 9x \rightarrow -16x + 25x - 5 = 9x - 5$

5) $5x + 5 - 15x \rightarrow 5x - 15x = -10x \rightarrow 5x + 5 - 15x = -10x + 5$

6) $4 + 5x - 6x - 5 \rightarrow 4 - 5 = -1, 5x - 6x = -x \rightarrow 4 + 5x - 6x - 5 = -x - 1$

7) $3x + 10 - 2x - 20 \rightarrow 10 - 20 = -10, 3x - 2x = x \rightarrow 3x + 10 - 2x - 20 = x - 10$

8) $-3 - 2x^2 - 5 + 3x \rightarrow -3 - 5 = -8 \rightarrow -3 - 2x^2 - 5 + 3x = -2x^2 + 3x - 8$

9) $-7 + 9x^2 - 2 + 2x \rightarrow -7 - 2 = -9 \rightarrow -7 + 9x^2 - 2 + 2x = 9x^2 + 2x - 9$

10) $4x^2 + 2x - 12x - 5 \rightarrow 2x - 12x = -10x \rightarrow 4x^2 + 2x - 12x - 5 = 4x^2 - 10x - 5$

11) $2x^2 - 3x - 5x + 6 - 9 \rightarrow -3x - 5x = -8x, 6 - 9 = -3 \rightarrow$

$2x^2 - 3x - 5x + 6 - 9 = 2x^2 - 8x - 3$

12) $x^2 - 6x - x + 2 - 3 \rightarrow -6x - x = -7x, 2 - 3 = -1 \rightarrow x^2 - 6x - x + 2 - 3 = x^2 - 7x - 1$

13) $10x^2 - x - 8x + 3 - 10 \rightarrow -x - 8x = -9x, 3 - 10 = -7 \rightarrow 10x^2 - x - 8x + 3 - 10 = 10x^2 - 9x - 7$

14) $4x^2 - 7x - x^2 + 2x + 5 \rightarrow 4x^2 - x^2 = 3x^2, -7x + 2x = -5x \rightarrow$

$4x^2 - 7x - x^2 + 2x + 5 = 3x^2 - 5x + 5$

15) $4x^2 + 3x^3 - x^2 + x \rightarrow 4x^2 - x^2 = 3x^2 \rightarrow 4x^2 + 3x^3 - x^2 + x = 3x^3 + 3x^2 + x$

16) $5x^4 + x^5 - x^4 + 4x^2 \rightarrow 5x^4 - x^4 = 4x^4 \rightarrow 5x^4 + x^5 - x^4 + 4x^2 =$

$4x^4 + x^5 + 4x^2 = x^5 + 4x^4 + 4x^2$

17) $15x^3 + 12x - 6x^2 - 9x^3 \rightarrow 15x^3 - 9x^3 = 6x^3 \rightarrow 15x^3 + 12x - 6x^2 - 9x^3 =$

$6x^3 + 12x - 6x^2 = 6x^3 - 6x^2 + 12x$

18) $(7x^3 - 2x^2) + (6x^2 - 13x) = 7x^3 - 2x^2 + 6x^2 - 13x \rightarrow -2x^2 + 6x^2 = 4x^2 \rightarrow$
$7x^3 - 2x^2 + 6x^2 - 13x = 7x^3 + 4x^2 - 13x$

19) $(9x^4 + 6x^3) + (11x^3 - 5x^4) = 9x^4 + 6x^3 + 11x^3 - 5x^4 \rightarrow 9x^4 - 5x^4 = 4x^4,$
$6x^3 + 11x^3 = 17x^3 \rightarrow 9x^4 + 6x^3 + 11x^3 - 5x^4 = 4x^4 + 17x^3$

20) $(15x^5 - 5x^3) - (4x^3 + 6x^2) = 15x^5 - 5x^3 - 4x^3 - 6x^2 \rightarrow -5x^3 - 4x^3 = -9x^3 \rightarrow$
$15x^5 - 5x^3 - 4x^3 - 6x^2 = 15x^5 - 9x^3 - 6x^2$

21) $(15x^4 + 7x^3) - (3x^3 - 26) = 15x^4 + 7x^3 - 3x^3 + 26 \rightarrow 7x^3 - 3x^3 = 4x^3 \rightarrow$
$15x^4 + 7x^3 - 3x^3 + 26 = 15x^4 + 4x^3 + 26$

22) $(22x^4 + 6x^3) - (-2x^3 - 4x^4) = 22x^4 + 6x^3 + 2x^3 + 4x^4 \rightarrow 22x^4 + 4x^4 =$
$26x^4, 6x^3 + 2x^3 = 8x^3 \rightarrow 22x^4 + 6x^3 + 2x^3 + 4x^4 = 26x^4 + 8x^3$

23) $(x^2 + 6x^3) + (-19x^2 + 6x^3) = x^2 + 6x^3 - 19x^2 + 6x^3 \rightarrow 6x^3 + 6x^3 = 12x^3,$
$-19x^2 + x^2 = -18x^2 \rightarrow x^2 + 6x^3 - 19x^2 + 6x^3 = 12x^3 - 18x^2$

24) $(2x^4 - x^3) + (-5x^3 - 7x^4) = 2x^4 - x^3 - 5x^3 - 7x^4 \rightarrow 2x^4 - 7x^4 =$

$-5x^4, -x^3 - 5x^3 = -6x^3 \rightarrow 2x^4 - x^3 - 5x^3 - 7x^4 = -5x^4 - 6x^3$

25) $3(5 + x) = (3) \times (5) + (3) \times x = 15 + 3x = 3x + 15$

26) $5(4 - x) = (5) \times (4) + (5) \times (-x) = 20 + (-5x) = -5x + 20$

27) $6(2 - 5x) = (6) \times (2) + (6) \times (-5x) = 12 + (-30x) = -30x + 12$

28) $(4 - 3x)7 = (4) \times (7) + (-3x) \times (7) = 28 + (-21x) = -21x + 28$

29) $8(3 - 3x) = (8) \times (3) + (8) \times (-3x) = 24 + (-24x) = -24x + 24$

30) $(-1)(-6 + 2x) = (-1) \times (-6) + (-1) \times (2x) = 6 + (-2x) = -2x + 6$

31) $(-5)(3x - 3) = (-5) \times (3x) + (-5) \times (-3) = -15x + 15$

32) $(-x + 10)(-3) = (-x) \times (-3) + (10) \times (-3) = 3x - 30$

33) $(-2)(2 - 6x) = (-2) \times (2) + (-2) \times (-6x) = -4 + 12x = 12x - 4$

34) $(-6x - 4)(-7) = (-6x) \times (-7) + (-4) \times (-7) = 42x + 28$

35) $x = 3 \rightarrow 12 - x = 12 - 3 = 9$

36) $x = 5 \rightarrow x + 7 = 5 + 7 = 12$

37) $x = 3 \rightarrow 3x - 5 = (3) \times (3) - 5 = 9 - 5 = 4$

38) $x = 2 \rightarrow 18 - 3x = 18 - (3) \times (2) = 18 - 6 = 12$

39) $x = 7 \rightarrow 5x - 4 = (5) \times (7) - 4 = 35 - 4 = 31$

40) $x = 6 \rightarrow 21 - x = 21 - 6 = 15$

41) $x = 5 \rightarrow 10x - 20 = (10) \times (5) - 20 = 50 - 20 = 30$

42) $x = -5 \rightarrow 4 - x = 4 - (-5) = 4 + 5 = 9$

43) $x = -2 \rightarrow 25 - 3x = 25 - (3) \times (-2) = 25 - (-6) = 25 + 6 = 31$

44) $x = -7 \rightarrow 16 - x = 16 - (-7) = 16 + 7 = 23$

45) $x = -13 \rightarrow 40 - 2x = 40 - (2) \times (-13) = 40 - (-26) = 40 + 26 = 66$

46) $x = -4 \rightarrow 20x - 6 = 20 \times (-4) - 6 = -80 - 6 = -86$

47) $x = -6 \rightarrow -11x - 19 = (-11) \times (-6) - 19 = 66 - 19 = 47$

48) $x = -8 \rightarrow -1 - 3x = (-1) - (3) \times (-8) = -1 - (-24) = -1 + 24 = 23$

49) $x = 3, \ y = 2 \rightarrow 3x + 2y = 3 \times (3) + 2(2) = 9 + 4 = 13$

50) $a = 4, \ b = 1 \rightarrow 2a - 6b = 2 \times (4) - 6(1) = 8 - 6 = 2$

51) $x = 5, \ y = 7 \rightarrow 2x - 4y - 5 = 2 \times (5) - 4(7) - 5 = 10 - 28 - 5 = -23$

52) $a = -3, \ b = 4 \rightarrow -3a + 4b + 2 = -3 \times (-3) + 4(4) + 2 = 9 + 16 + 2 = 27$

53) $x = -4, \ y = -3 \rightarrow 2x - 6 - 4y = 2 \times (-4) - 6 - 4(-3) = -8 - 6 + 12 = -2$

6 Equations and Inequalities

Math topics that you'll learn in this chapter:

1. One-Step Equations
2. Multi-Step Equations
3. System of Equations
4. Graphing Single–Variable Inequalities
5. One-Step Inequalities
6. Multi-Step Inequalities

67

One–Step Equations

☆ The values of two expressions on both sides of an equation are equal. Example: $ax = b$. In this equation, ax is equal to b.

☆ Solving an equation means finding the value of the variable.

☆ You only need to perform one Math operation to solve the one-step equations.

☆ To solve a one-step equation, find the inverse (opposite) operation is being performed.

☆ The inverse operations are:

❖ Addition and subtraction

❖ Multiplication and division

Examples:

Example 1. Solve this equation for x. $6x = 18 \rightarrow x = ?$

Solution: Here, the operation is multiplication (variable x is multiplied by 6) and its inverse operation is division. To solve this equation, divide both sides of equation by 6: $6x = 18 \rightarrow \frac{6x}{6} = \frac{18}{6} \rightarrow x = 3$

Example 2. Solve this equation. $x + 5 = 0 \rightarrow x = ?$

Solution: In this equation, 5 is added to the variable x. The inverse operation of addition is subtraction. To solve this equation, subtract 5 from both sides of the equation: $x + 5 - 5 = 0 - 5$. Then: $x + 5 - 5 = 0 - 5 \rightarrow x = -5$

Example 3. Solve this equation for x. $x - 11 = 0$

Solution: Here, the operation is subtraction and its inverse operation is addition. To solve this equation, add 11 to both sides of the equation: $x - 11 + 11 = 0 + 11 \rightarrow x = 11$

Multi–Step Equations

☆ To solve a multi-step equation, combine "like" terms on one side.

☆ Bring variables to one side by adding or subtracting.

☆ Simplify using the inverse of addition or subtraction.

☆ Simplify further by using the inverse of multiplication or division.

☆ Check your solution by plugging the value of the variable into the original equation.

Examples:

Example 1. Solve this equation for x. $5x - 6 = 26 - 3x$

Solution: First, bring variables to one side by adding $3x$ to both sides. Then:
$5x - 6 + 3x = 26 - 3x + 3x \rightarrow 5x - 6 + 3x = 26$.
Simplify: $8x - 6 = 26$. Now, add 6 to both sides of the equation:
$8x - 6 + 6 = 26 + 6 \rightarrow 8x = 32 \rightarrow$ Divide both sides by 8:
$8x = 32 \rightarrow \dfrac{8x}{8} = \dfrac{32}{8} \rightarrow x = 4$

Let's check this solution by substituting the value of 4 for x in the original equation:
$\quad x = 4 \rightarrow 5x - 6 = 26 - 3x \rightarrow 5(4) - 6 = 26 - 3(4) \rightarrow 20 - 6 = 26 - 12 \rightarrow 14 = 14$
The answer $x = 4$ is correct.

Example 2. Solve this equation for x. $6x - 3 = 15$

Solution: Add 3 to both sides of the equation.
$6x - 3 = 15 \rightarrow 6x - 3 + 3 = 15 + 3 \rightarrow 6x = 18$
Divide both sides by 6, then: $6x = 18 \rightarrow \frac{6x}{6} = \frac{18}{6} \rightarrow x = 3$

Now, check the solution:
$x = 4 \rightarrow 6(3) - 3 = 15 \rightarrow 18 - 3 = 15 \rightarrow 15 = 15$
The answer $x = 4$ is correct.

System of Equations

☆ A system of equations contains two equations and two variables. For example, consider the system of equations: $x - y = 1$ and $x + y = 5$

☆ The easiest way to solve a system of equations is using the elimination method. The elimination method uses the addition property of equality. You can add the same value to each side of an equation.

☆ For the first equation above, you can add $x + y$ to the left side and 5 to the right side of the first equation: $x - y + (x + y) = 1 + 5$. Now, if you simplify, you get: $x - y + (x + y) = 1 + 5 \rightarrow 2x = 6 \rightarrow x = 3$. Now, substitute 3 for the x in the first equation: $3 - y = 1$. By solving this equation, $y = 2$

Example:

What is the value of $x + y$ in this system of equations?

$$\begin{cases} -x + y = 18 \\ 2x - 6y = -12 \end{cases}$$

Solution: Solving the system of equations by elimination:
Multiply the first equation by (2), then add it to the second equation.

$$\begin{matrix} 2(-x + y = 18) \\ 2x - 6y = -12 \end{matrix} \Rightarrow \begin{matrix} -2x + 2y = 36 \\ 2x - 6y = -12 \end{matrix} \Rightarrow (-2x) + 2x + 2y - 6y = 36 - 12 \Rightarrow -4y = 24 \Rightarrow$$

$y = -6$

Plug in the value of y into one of the equations and solve for x.
$-x + (-6) = 18 \Rightarrow -x - 6 = 18 \Rightarrow -x = 24 \Rightarrow x = -24$
Thus, $x + y = -24 - 6 = -30$

Graphing Single–Variable Inequalities

☆ An inequality compares two expressions using an inequality sign.

☆ Inequality signs are: "less than" <, "greater than" >, "less than or equal to" ≤, and "greater than or equal to" ≥.

☆ To graph a single–variable inequality, find the value of the inequality on the number line.

☆ For less than (<) or greater than (>) draw an open circle on the value of the variable. If there is an equal sign too, then use a filled circle.

☆ Draw an arrow to the right for greater or to the left for less than.

Examples:

Example 1. Draw a graph for this inequality. $x > 3$

Solution: Since the variable is greater than 3, then we need to find 3 in the number line and draw an open circle on it. Then, draw an arrow to the right.

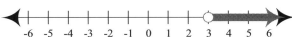

Example 2. Graph this inequality. $x \leq -1$.

Solution: Since the variable is less than or equal to −1, then we need to find −1 on the number line and draw a filled circle on it. Then, draw an arrow to the left.

One–Step Inequalities

☆ An inequality compares two expressions using an inequality sign.

☆ Inequality signs are: "less than" <, "greater than" >, "less than or equal to" ≤, and "greater than or equal to" ≥.

☆ You only need to perform one Math operation to solve the one-step inequalities.

☆ To solve one-step inequalities, find the inverse (opposite) operation is being performed.

☆ For dividing or multiplying both sides by negative numbers, flip the direction of the inequality sign.

Examples:

Example 1. Solve this inequality for x. $x + 7 \geq 2$

Solution: The inverse (opposite) operation of addition is subtraction. In this inequality, 7 is added to x. To isolate x we need to subtract 7 from both sides of the inequality.
Then: $x + 7 \geq 2 \rightarrow x + 7 - 7 \geq 2 - 7 \rightarrow x \geq -5$. The solution is: $x \geq -5$

Example 2. Solve the inequality. $x - 2 > -12$

Solution: 2 is subtracted from x. Add 2 to both sides.
$x - 2 > -12 \rightarrow x - 2 + 2 > -12 + 2 \rightarrow x > -10$

Example 3. Solve. $6x \leq -36$

Solution: 6 is multiplied to x. Divide both sides by 6.
Then: $6x \leq -36 \rightarrow \frac{6x}{6} \leq \frac{-36}{6} \rightarrow x \leq -6$

Example 4. Solve. $-2x \leq 10$

Solution: -2 is multiplied to x. Divide both sides by -2. Remember when dividing or multiplying both sides of an inequality by negative numbers, flip the direction of the inequality sign.
Then: $-2x \leq 10 \rightarrow \frac{-2x}{-2} \geq \frac{10}{-2} \rightarrow x \geq -5$

Multi–Step Inequalities

☆ To solve a multi-step inequality, combine "like" terms on one side.

☆ Bring variables to one side by adding or subtracting.

☆ Isolate the variable.

☆ Simplify using the inverse of addition or subtraction.

☆ Simplify further by using the inverse of multiplication or division.

☆ For dividing or multiplying both sides by negative numbers, flip the direction of the inequality sign.

Examples:

Example 1. Solve this inequality. $4x - 1 \leq 23$

Solution: In this inequality, 1 is subtracted from $4x$. The inverse of subtraction is addition. Add 1 to both sides of the inequality:
$4x - 1 + 1 \leq 23 + 1 \rightarrow 4x \leq 24$
Now, divide both sides by 4. Then: $4x \leq 24 \rightarrow \frac{4x}{4} \leq \frac{24}{4} \rightarrow x \leq 6$
The solution of this inequality is $x \leq 6$.

Example 2. Solve this inequality. $2x - 6 < 18$

Solution: First, add 6 to both sides: $2x - 6 + 6 < 18 + 6$
Then simplify: $2x - 6 + 6 < 18 + 6 \rightarrow 2x < 24$
Now divide both sides by 2: $\frac{2x}{2} < \frac{24}{2} \rightarrow x < 12$

Example 3. Solve this inequality. $-4x - 8 \geq 12$

Solution: First, add 8 to both sides:
$-4x - 8 + 8 \geq 12 + 8 \rightarrow -4x \geq 20$
Divide both sides by -4. Remember that you need to flip the direction of inequality sign. $-4x \geq 20 \rightarrow \frac{-4x}{-4} \leq \frac{20}{-4} \rightarrow x \leq -5$

Day 6: Practices

✒ Solve each equation. (One–Step Equations)

1) $x + 2 = 5 \rightarrow x =$

2) $8 = 13 + x \rightarrow x =$

3) $-6 = 7 + x \rightarrow x =$

4) $x - 5 = -3 \rightarrow x =$

5) $-13 = x - 15 \rightarrow x =$

6) $-10 + x = -4 \rightarrow x =$

7) $-19 + x = 7 \rightarrow x =$

8) $-6x = 24 \rightarrow x =$

9) $\frac{x}{4} = -5 \rightarrow x =$

10) $-2x = -4 \rightarrow x =$

✒ Solve each equation. (Multi–Step Equations)

11) $2(x + 5) = 16 \rightarrow x =$

12) $-6(3 - x) = 18 \rightarrow x =$

13) $25 = -5(x + 4) \rightarrow x =$

14) $-12 = 6(9 + x) \rightarrow x =$

15) $11(x + 5) = -22 \rightarrow x =$

16) $-27 - 36x = 45 \rightarrow x =$

17) $3x - 4 = x - 12 \rightarrow x =$

18) $-8x + x - 11 = 24 \rightarrow x =$

✒ Solve each system of equations.

19) $\begin{cases} x + 4y = 29 \\ x + 2y = 5 \end{cases}$ $x = \underline{}$ $y = \underline{}$

20) $\begin{cases} 2x + y = 36 \\ x + 4y = 4 \end{cases}$ $x = \underline{}$ $y = \underline{}$

21) $\begin{cases} 2x + 5y = 15 \\ x + y = 6 \end{cases}$ $x = \underline{}$ $y = \underline{}$

22) $\begin{cases} 2x - 2y = -16 \\ -9x + 2y = -19 \end{cases}$ $x = \underline{}$ $y = \underline{}$

✍ Draw a graph for each inequality.

23) $x \leq 1$

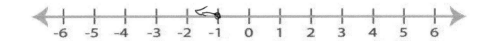

24) $x > -4$

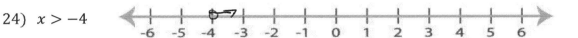

✍ Solve each inequality and graph it.

25) $x - 3 \geq -2$

26) $7x - 6 < 8$

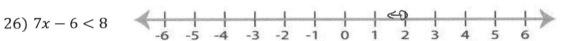

✍ Solve each inequality.

27) $x + 11 > 3$

28) $x + 4 > 1$

29) $-6 + 3x \leq 21$

30) $-5 + 4x \leq 19$

31) $4 + 9x \leq 31$

32) $8(x + 3) \geq -16$

33) $3(6 + x) \geq 18$

34) $3(x - 2) < -9$

35) $15 + 9x < -30$

36) $3(6 - x) \geq -27$

37) $4(x - 5) \geq -32$

38) $6(x + 4) < -24$

39) $7(x - 8) \geq -49$

40) $-(-6 - 5x) > -39$

41) $2(1 - 2x) > -66$

42) $-3(3 - 2x) > -33$

Day 6: Answers

1) $x + 2 = 5 \rightarrow x = 5 - 2 = 3$

2) $8 = 13 + x \rightarrow x = 8 - 13 = -5$

3) $-6 = 7 + x \rightarrow x = -6 - 7 = -13$

4) $x - 5 = -3 \rightarrow x = -3 + 5 = 2$

5) $-13 = x - 15 \rightarrow x = -13 + 15 = 2$

6) $-10 + x = -4 \rightarrow x = -4 + 10 = 6$

7) $-19 + x = 7 \rightarrow x = 7 + 19 = 26$

8) $-6x = 24 \rightarrow x = \frac{24}{-6} = -4$

9) $\frac{x}{4} = -5 \rightarrow x = -5 \times 4 = -20$

10) $-2x = -4 \rightarrow x = \frac{-4}{-2} = 2$

11) $2(x + 5) = 16 \rightarrow \frac{2(x+5)}{2} = \frac{16}{2} \rightarrow (x + 5) = 8 \rightarrow x = 8 - 5 = 3$

12) $-6(3 - x) = 18 \rightarrow \frac{-6(3-x)}{-6} = \frac{18}{-6} \rightarrow (3 - x) = -3 \rightarrow x = 3 + 3 = 6$

13) $25 = -5(x + 4) \rightarrow \frac{25}{-5} = \frac{-5(x+4)}{-5} \rightarrow -5 = (x + 4) \rightarrow x = -5 - 4 = -9$

14) $-12 = 6(9 + x) \rightarrow \frac{-12}{6} = \frac{6(9+x)}{6} \rightarrow -2 = (9 + x) \rightarrow x = -2 - 9 = -11$

15) $11(x + 5) = -22 \rightarrow \frac{11(x+5)}{11} = \frac{-22}{11} \rightarrow (x + 5) = -2 \rightarrow x = -2 - 5 = -7$

16) $-27 - 36x = 45 \rightarrow -36x = 45 + 27 \rightarrow -36x = 72 \rightarrow \frac{-36x}{-2} = \frac{72}{-2} \rightarrow x = -2$

17) $3x - 4 = x - 12 \rightarrow 3x - 4 - x = x - 12 - x \rightarrow 2x - 4 = -12 \rightarrow 2x = -12 + 4 \rightarrow$
$2x = -8 \rightarrow \frac{2x}{2} = \frac{-8}{2} \rightarrow x = -4$

18) $-8x + x - 11 = 24 \rightarrow -7x = 24 + 11 \rightarrow \frac{-7x}{-7} = \frac{35}{-7} \rightarrow x = -5$

19) $\begin{cases} x + 4y = 29 \\ x + 2y = 5 \end{cases} \rightarrow \begin{matrix} -(x + 4y = 29) \\ x + 2y = 5 \end{matrix} \rightarrow \begin{matrix} -x - 4y = -29 \\ x + 2y = 5 \end{matrix} \rightarrow (-x) + x - 4y + 2y =$
$-29 + 5 \rightarrow -2y = -24 \rightarrow \frac{-2y}{-2} = \frac{-24}{-2} \rightarrow y = 12$

Plug in the value of y into one of the equations and solve for x.
$x + 2y = 5 \rightarrow x + 2(12) = 5 \rightarrow x + 24 = 5 \rightarrow x = 5 - 24 = -19$

20) $\begin{cases} 2x + y = 36 \\ x + 4y = 4 \end{cases} \rightarrow \begin{matrix} 2x + y = 36 \\ -2(x + 4y = 4) \end{matrix} \rightarrow \begin{matrix} 2x + y = 36 \\ -2x - 8y = -8 \end{matrix} \rightarrow 2x - 2x + y - 8y = 36 -$

$8 \rightarrow -7y = 28 \rightarrow \frac{-7y}{-7} = \frac{28}{-7} \rightarrow y = -4$

Plug in the value of y into one of the equations and solve for x.

$x + 4y = 4 \rightarrow x + 4(-4) = 4 \rightarrow x - 16 = 4 \rightarrow x = 4 + 16 = 20$

21) $\begin{cases} 2x + 5y = 15 \\ x + y = 6 \end{cases} \rightarrow \begin{matrix} 2x + 5y = 15 \\ -2(x + y = 6) \end{matrix} \rightarrow \begin{matrix} 2x + 5y = 15 \\ -2x - 2y = -12 \end{matrix} \rightarrow 2x - 2x + 5y - 2y = 15 -$

$12 \rightarrow 3y = 3 \rightarrow \frac{3y}{3} = \frac{3}{3} \rightarrow y = 1$

Plug in the value of y into one of the equations and solve for x.

$x + y = 6 \rightarrow x + 1 = 6 \rightarrow x = 6 - 1 = 5$

22) $\begin{cases} 2x - 2y = -16 \\ -9x + 2y = -19 \end{cases} \rightarrow 2x - 9x - 2y + 2y = -16 - 19 \rightarrow -7x = -35 \rightarrow \frac{-7x}{-7} =$

$\frac{-35}{-7} \rightarrow x = 5$

Plug in the value of x into one of the equations and solve for y.

$2x - 2y = -16 \rightarrow 2(5) - 2y = -16 \rightarrow 10 - 2y = -16 \rightarrow 10 + 16 = 2y \rightarrow 26 = 2y$

$\rightarrow \frac{26}{2} = \frac{2y}{2} \rightarrow y = 13$

23) $x \leq 1$

24) $x > -4$

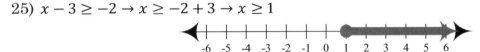

25) $x - 3 \geq -2 \rightarrow x \geq -2 + 3 \rightarrow x \geq 1$

26) $7x - 6 < 8 \rightarrow 7x < 8 + 6 \rightarrow 7x < 14 \rightarrow x < \frac{14 \div 7}{7 \div 7} \rightarrow x < 2$

27) $x + 11 > 3 \rightarrow x > 3 - 11 \rightarrow x > -8$

28) $x + 4 > 1 \rightarrow x > 1 - 4 \rightarrow x > -3$

29) $-6 + 3x \leq 21 \rightarrow 3x \leq 21 + 6 \rightarrow 3x \leq 27 \rightarrow x \leq \frac{27 \div 3}{3 \div 3} \rightarrow x \leq 9$

30) $-5 + 4x \leq 19 \rightarrow 4x \leq 19 + 5 \rightarrow 4x \leq 24 \rightarrow x \leq \frac{24 \div 4}{4 \div 4} \rightarrow x \leq 6$

31) $4 + 9x \leq 31 \rightarrow 9x \leq 31 - 4 \rightarrow 9x \leq 27 \rightarrow x \leq \frac{27 \div 9}{9 \div 9} \rightarrow x \leq 3$

32) $8(x + 3) \geq -16 \rightarrow x + 3 \geq \frac{-16}{8} \rightarrow x + 3 \geq -2 \rightarrow x \geq -2 - 3 \rightarrow x \geq -5$

33) $3(6 + x) \geq 18 \rightarrow 6 + x \geq \frac{18}{3} \rightarrow 6 + x \geq 6 \rightarrow x \geq 6 - 6 \rightarrow x \geq 0$

34) $3(x - 2) < -9 \rightarrow x - 2 < \frac{-9}{3} \rightarrow x - 2 < -3 \rightarrow x < -3 + 2 \rightarrow x < -1$

35) $15 + 9x < -30 \rightarrow 9x < -30 - 15 \rightarrow 9x < -45 \rightarrow x < \frac{-45}{9} \rightarrow x < -5$

36) $3(6 - x) \geq -27 \rightarrow 6 - x \geq \frac{-27}{3} \rightarrow 6 - x \geq -9 \rightarrow -x \geq -9 - 6 \rightarrow \frac{-x}{-1} \leq \frac{-15}{-1} \rightarrow$

$x \leq 15$

37) $4(x - 5) \geq -32 \rightarrow x - 5 \geq \frac{-32}{4} \rightarrow x - 5 \geq -8 \rightarrow x \geq -8 + 5 \rightarrow x \geq -3$

38) $6(x + 4) < -24 \rightarrow x + 4 < \frac{-24}{6} \rightarrow x + 4 < -4 \rightarrow x < -4 - 4 \rightarrow x < -8$

39) $7(x - 8) \geq -49 \rightarrow x - 8 \geq \frac{-49}{7} \rightarrow x - 8 \geq -7 \rightarrow x \geq -7 + 8 \rightarrow x \geq 1$

40) $-(-6 - 5x) > -39 \rightarrow 6 + 5x > -39 \rightarrow 5x > -39 - 6 \rightarrow 5x > -45 \rightarrow$

$x > \frac{-45}{5} \rightarrow x > -9$

41) $2(1 - 2x) > -66 \rightarrow 1 - 2x > \frac{-66}{2} \rightarrow 1 - 2x > -33 \rightarrow -2x > -33 - 1 \rightarrow$

$-2x > -34 \rightarrow x < \frac{-34}{-2} \rightarrow x < \frac{-34}{-2} \rightarrow x < 17$

42) $-3(3 - 2x) > -33 \rightarrow -9 + 6x > -33 \rightarrow 6x > -33 + 9 \rightarrow 6x > -24 \rightarrow$

$\frac{6x}{6} > \frac{-24}{6} \rightarrow x > -4$

7 Lines and Slope

Math topics that you'll learn in this chapter:

1. Finding Slope
2. Graphing Lines Using Slope–Intercept Form
3. Writing Linear Equations
4. Finding Midpoint
5. Finding Distance of Two Points
6. Graphing Linear Inequalities

Finding Slope

☆ The slope of a line represents the direction of a line on the coordinate plane.

☆ A coordinate plane contains two perpendicular number lines. The horizontal line is x and the vertical line is y. The point at which the two axes intersect is called the origin. An ordered pair (x, y) shows the location of a point.

☆ A line on a coordinate plane can be drawn by connecting two points.

☆ To find the slope of a line, we need the equation of the line or two points on the line.

☆ The slope of a line with two points A (x_1, y_1) and B (x_2, y_2) can be found by using this formula: $\frac{y_2 - y_1}{x_2 - x_1} = \frac{rise}{run}$

☆ The equation of a line is typically written as $y = mx + b$ where m is the slope and b is the y-intercept.

Examples:

Example 1. Find the slope of the line through these two points:

A$(5, -5)$ *and* B$(9, 7)$.

Solution: Slope $= \frac{y_2 - y_1}{x_2 - x_1}$. Let (x_1, y_1) be A$(5, -5)$ and (x_2, y_2) be B$(9, 7)$.

(Remember, you can choose any point for (x_1, y_1) and (x_2, y_2)).

Then: slope $= \frac{y_2 - y_1}{x_2 - x_1} = \frac{7 - (-5)}{9 - 5} = \frac{12}{4} = 3$

The slope of the line through these two points is 3.

Example 2. Find the slope of the line with equation $y = -2x + 8$

Solution: When the equation of a line is written in the form of $y = mx + b$, the slope is m. In this line: $y = -2x + 8$, the slope is -2.

Graphing Lines Using Slope–Intercept Form

✮ Slope–intercept form of a line: given the slope **m** and the **y**–intercept (the intersection of the line and y-axis) **b**, then the equation of the line is:

$$y = mx + b$$

✮ To draw the graph of a linear equation in a slope-intercept form on the xy coordinate plane, find two points on the line by plugging two values for x and calculating the values of y.

✮ You can also use the slope (m) and one point to graph the line.

Example:

Sketch the graph of $y = -3x + 6$.

Solution: To graph this line, we need to find two points. When x is zero the value of y is 6. And when y is 0 the value of x is 2.

$$x = 0 \rightarrow y = -3(0) + 6 = 6$$
$$y = 0 \rightarrow 0 = -3x + 6 \rightarrow x = 2$$

Now, we have two points:
$(0, 6)$ and $(2, 0)$.
Find the points on the coordinate plane and graph the line. Remember that the slope of the line is -3.

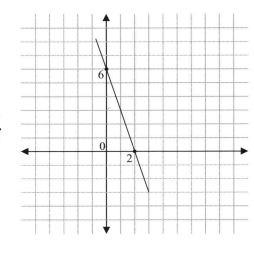

Writing Linear Equations

☆ The equation of a line in slope-intercept form: $y = mx + b$

☆ To write the equation of a line, first identify the slope.

☆ Find the y-intercept. This can be done by substituting the slope and the coordinates of a point (x, y) on the line.

Examples:

Example 1. What is the equation of the line that passes through $(-7, 2)$ and has a slope of 4?

Solution: The general slope-intercept form of the equation of a line is $y = mx + b$, where m is the slope and b is the y-intercept.
By substitution of the given point and given slope:
$y = mx + b \rightarrow 2 = (4)(-7) + b$. So, $b = 2 + 28 = 30$, and the required equation of the line is: $y = 4x + 30$

Example 2. Write the equation of the line through two points $A(5, 2)$ and $B(3, -4)$.

Solution: First, find the slope: $Slop = \frac{y_2 - y_1}{x_2 - x_1} = \frac{-4 - 2}{3 - 5} = \frac{-6}{-2} = 3 \rightarrow m = 3$

To find the value of b, use either points and plug in the values of x and y in the equation. The answer will be the same: $y = x + b$. Let's check both points. Then:
$(5, 2) \rightarrow y = mx + b \rightarrow 2 = 3(5) + b \rightarrow b = -13$
$(3, -4) \rightarrow y = mx + b \rightarrow -4 = 3(3) + b \rightarrow b = -13$.
The y-intercept of the line is -13. The equation of the line is: $y = 3x - 13$

Example 3. What is the equation of the line that passes through $(3, -4)$ and has a slope of 2?

Solution: The general slope-intercept form of the equation of a line is $y = mx + b$, where m is the slope and b is the y-intercept. By substitution of the given point and given slope: $y = mx + b \rightarrow -4 = (2)(3) + b$
So, $b = -4 - 6 = -10$, and the equation of the line is: $y = 2x - 10$.

Finding Midpoint

☆ The middle of a line segment is its midpoint.

☆ The Midpoint of two endpoints A (x_1, y_1) and B (x_2, y_2) can be found using this formula: $M = \left(\frac{x_1 + x_2}{2}, \frac{y_1 + y_2}{2}\right)$

Examples:

Example 1. Find the midpoint of the line segment with the given endpoints. $(3, 5), (1, 3)$

Solution: Midpoint $= \left(\frac{x_1 + x_2}{2}, \frac{y_1 + y_2}{2}\right) \rightarrow (x_1, y_1) = (3, 5)$ and $(x_2, y_2) = (1, 3)$

Midpoint $= \left(\frac{3+1}{2}, \frac{5+3}{2}\right) \rightarrow \left(\frac{4}{2}, \frac{8}{2}\right) \rightarrow M(2, 4)$

Example 2. Find the midpoint of the line segment with the given endpoints. $(-1, 3), (9, -9)$

Solution: Midpoint $= \left(\frac{x_1 + x_2}{2}, \frac{y_1 + y_2}{2}\right) \rightarrow (x_1, y_1) = (-1, 3)$ and $(x_2, y_2) = (9, -9)$

Midpoint $= \left(\frac{-1+9}{2}, \frac{3+(-9)}{2}\right) \rightarrow \left(\frac{8}{2}, \frac{-6}{2}\right) \rightarrow M(4, -3)$

Example 3. Find the midpoint of the line segment with the given endpoints. $(8, 4), (-2, 6)$

Solution: Midpoint $= \left(\frac{x_1 + x_2}{2}, \frac{y_1 + y_2}{2}\right) \rightarrow (x_1, y_1) = (8, 4)$ and $(x_2, y_2) = (-2, 6)$

Midpoint $= \left(\frac{8-2}{2}, \frac{4+6}{2}\right) \rightarrow \left(\frac{6}{2}, \frac{10}{2}\right) \rightarrow M(3, 5)$

Example 4. Find the midpoint of the line segment with the given endpoints. $(7, -4), (-3, -8)$

Solution: Midpoint $= \left(\frac{x_1 + x_2}{2}, \frac{y_1 + y_2}{2}\right) \rightarrow (x_1, y_1) = (7, -4)$ and $(x_2, y_2) = (-3, -8)$

Midpoint $= \left(\frac{7-3}{2}, \frac{-4-8}{2}\right) \rightarrow \left(\frac{4}{2}, \frac{-12}{2}\right) \rightarrow M(2, -6)$

Finding Distance of Two Points

☆ Use the following formula to find the distance of two points with the
   coordinates A (x_1, y_1) and B (x_2, y_2):

$$d = \sqrt{(x_2 - x_1)^2 + (y_2 - y_1)^2}$$

Examples:

Example 1. Find the distance between $(5, -6)$ and $(-3, 9)$. on the coordinate
plane.

Solution: Use distance of two points formula: $d = \sqrt{(x_2 - x_1)^2 + (y_2 - y_1)^2}$
$(x_1, y_1) = (5, -6)$ and $(x_2, y_2) = (-3, 9)$. Then: $d = \sqrt{(x_2 - x_1)^2 + (y_2 - y_1)^2} =$
$\sqrt{(-3 - 5)^2 + (9 - (-6))^2} = \sqrt{(-8)^2 + (15)^2} = \sqrt{64 + 225} = \sqrt{289} = 17$.
Then: $d = 17$

Example 2. Find the distance of two points $(-3, 10)$ and $(-9, 2)$

Solution: Use distance of two points formula: $d = \sqrt{(x_2 - x_1)^2 + (y_2 - y_1)^2}$
$(x_1, y_1) = (-3, 10)$ and $(x_2, y_2) = (-9, 2)$
Then: $d = \sqrt{(x_2 - x_1)^2 + (y_2 - y_1)^2} \rightarrow d = \sqrt{(-9 - (-3))^2 + (2 - 10)^2} =$
$\sqrt{(-6)^2 + (-8)^2} = \sqrt{36 + 64} = \sqrt{100} = 10$. Then: $d = 10$

Example 3. Find the distance between $(-8, 7)$ and $(4, -9)$.

Solution: Use distance of two points formula: $d = \sqrt{(x_2 - x_1)^2 + (y_2 - y_1)^2}$
$(x_1, y_1) = (-8, 7)$ and $(x_2, y_2) = (4, -9)$. Then: $d = \sqrt{(x_2 - x_1)^2 + (y_2 - y_1)^2}$
$d = \sqrt{(4 - (-8))^2 + (-9 - 7)^2} = \sqrt{(12)^2 + (-16)^2} = \sqrt{144 + 256} = \sqrt{400} = 20$.
Then: $d = 20$

Graphing Linear Inequalities

☆ To graph a linear inequality, first draw a graph of the "equals" line.

☆ Use a dash line for less than ($<$) and greater than ($>$) signs and a solid line for less than and equal to ($\leq$) and greater than and equal to ($\geq$).

☆ Choose a testing point. (it can be any point on both sides of the line.)

☆ Put the value of (x, y) of that point in the inequality. If that works, that part of the line is the solution. If the values don't work, then the other part of the line is the solution.

Example:

Sketch the graph of inequality: $y > 2x - 5$

Solution: To draw the graph of $y > 2x - 5$, you first need to graph the line:

$y = 2x - 5$

Since there is a greater than ($>$) sign, draw a dash line.

The slope is 2 and y-intercept is -5.

Then, choose a testing point and substitute the value of x and y from that point into the inequality. The easiest point to test is the origin: $(0, 0)$

$(0, 0) \rightarrow y > 2x - 5 \rightarrow 0 > 2(0) - 5 \rightarrow 0 > -5$

This is correct! 0 is greater than -5. So, this part of the line (on the left side) is the solution of this inequality.

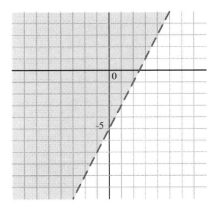

Day 7: Practices

✍ Find the slope of each line.

1) $y = x - 3$

2) $y = 3x + 4$

3) $y = -2x + 4$

4) Line through $(2, 5)$ and $(3, -4)$

5) Line through $(0, 6)$ and $(2, 4)$

6) Line through $(-2, 4)$ and $(3, -6)$

✍ Sketch the graph of each line. (Using Slope–Intercept Form)

7) $y = x + 3$

8) $y = x - 3$

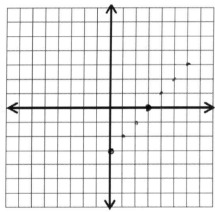

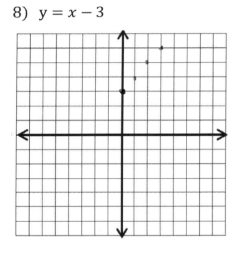

✍ Solve.

9) What is the equation of a line with slope 3 and intercept 12?

_____

10) What is the equation of a line with slope 4 and passes through point $(2, 4)$? _____

11) What is the equation of a line with slope -2 and passes through point $(5, -3)$?

_____

12) The slope of a line is -5 and it passes through point $(-4, 3)$. What is the equation of the line? _____

13) The slope of a line is -6 and it passes through point $(-2, -3)$. What is the equation of the line? _____

✍ Sketch the graph of each linear inequality.

14) $y > 3x - 3$

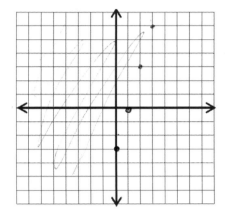

15) $y > -2x + 1$

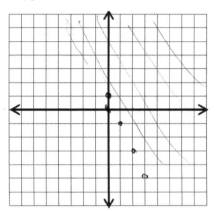

✍ Find the midpoint of the line segment with the given endpoints.

16) $(4, 1), (2, 3)$

17) $(3, 6), (5, 4)$

18) $(7, 1), (1, 3)$

19) $(2, 8), (2, 10)$

20) $(3, -2), (-1, 6)$

21) $(-1, -3), (1, 5)$

22) $(1, 4), (-7, 6)$

23) $(-3, 5), (7, -9)$

✍ Find the distance between each pair of points.

24) $(-8, -1), (-4, 2)$

25) $(-15, 2), (5, -13)$

26) $(-1, 11), (-7, 3)$

27) $(0, 11), (9, 11)$

28) $(-2, 4), (3, -8)$

29) $(6, -7), (-9, 1)$

30) $(8, -4), (-4, -20)$

31) $(5, 1), (9, -2)$

32) $(-8, -17), (2, 7)$

33) $(18, 21), (-12, 5)$

Day 7: Answers

1) $y = mx + b$, the slope is m. In this line: $y = x - 3$, the slope is $m = 1$.

2) $y = 3x + 4$, the slope is $m = 3$.

3) $y = -2x + 4$, the slope is $m = -2$.

4) $(x_1, y_1) = (2, 5)$ and $(x_2, y_2) = (3, 4) \rightarrow m = \frac{y_2 - y_1}{x_2 - x_1} = \frac{4-5}{3-2} = \frac{-1}{1} = -1$

5) $(x_1, y_1) = (0, 6)$ and $(x_2, y_2) = (2, -4) \rightarrow m = \frac{y_2 - y_1}{x_2 - x_1} = \frac{-4-6}{2-0} = \frac{-10}{2} = -5$

6) $(x_1, y_1) = (-2, 4)$ and $(x_2, y_2) = (3, -6) \rightarrow m = \frac{y_2 - y_1}{x_2 - x_1} = \frac{-6-4}{3-(-2)} = \frac{-6-4}{3+2} = \frac{-10}{5} = -2$

7) $y = x + 3$

    $x = 0 \rightarrow y = 0 + 3 = 3 \rightarrow (0, 3)$

    $y = 0 \rightarrow 0 = x + 3 \rightarrow x = -3$

        $\rightarrow (-3, 0)$

    $x = 1 \rightarrow y = 1 + 3 = 4 \rightarrow (1, 4)$

      $y = 1 \rightarrow 1 = x + 3 \rightarrow x = 1 - 3$

        $= -2 \rightarrow (-2, 1)$

8) $y = x - 3$

    $x = 0 \rightarrow y = 0 - 3 = -3 \rightarrow (0, -3)$

    $y = 0 \rightarrow 0 = x - 3 \rightarrow x = 3 \rightarrow$

        $\rightarrow (3, 0)$

    $x = 1 \rightarrow y = 1 - 3 = -2 \rightarrow (1, -2)$

    $y = 1 \rightarrow 1 = x - 3 \rightarrow x = 1 + 3 = 4$

        $\rightarrow (4, 1)$

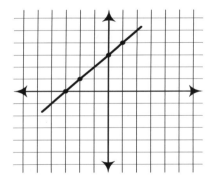

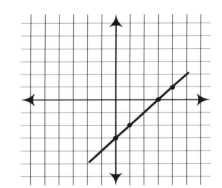

9) The general slope-intercept form of the equation of a line is $y = mx + b$, where m is the slope and b is the y-intercept $\rightarrow y = 3x + 12$

10) $y = mx + b \rightarrow 4 = 4(2) + b \rightarrow 4 = 8 + b \rightarrow b = 4 - 8 = -4 \rightarrow y = 4x - 4$

11) $y = mx + b \rightarrow -3 = -2(5) + b \rightarrow -3 = -10 + b \rightarrow b = -3 + 10 = 7 \rightarrow y = -2x + 7$

12) $y = mx + b \rightarrow 3 = -5(-4) + b \rightarrow 3 = 20 + b \rightarrow b = 3 - 20 = -17 \rightarrow$
    $y = -5x - 17$

13) $y = mx + b \rightarrow -3 = -6(-2) + b \rightarrow -3 = 12 + b \rightarrow b = -3 - 12 = -15 \rightarrow$
    $y = -6x - 15$

14) $y > 3x - 3$

    $x = 0 \rightarrow y = 0 - 3 = -3 \rightarrow (0, -3)$

$y = 0 \rightarrow 0 = 3x - 3 \rightarrow 3x = 3 \rightarrow x = 1 \rightarrow (1, 0)$

The easiest point to test is the origin: $(0, 0)$

$(0,0) \rightarrow y > 3x - 3 \rightarrow 0 > 3(0) - 3 \rightarrow 0 > -3$

This is correct! 0 is greater than -3. So, this part of the line (on the left side) is the solution of this inequality.

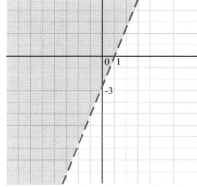

15) $y > -2x + 1$

$\quad x = 0 \rightarrow y = 0 + 1 = 1 \rightarrow (0,1)$

$\quad y = 0 \rightarrow 0 = -2x + 1 \rightarrow -2x = -1 \rightarrow x = \dfrac{-1}{-2}$

$\qquad\qquad = 0.5 \rightarrow (0.5, 0)$

The easiest point to test is the origin: $(0, 0)$

$(0,0) \rightarrow y > -2x + 1 \rightarrow 0 > -2(0) + 1 \rightarrow 0 > 1$

This is incorrect! 0 is less than 1. So, this part of the line (on the right side) is the solution of this inequality.

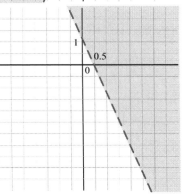

16) $M = \left(\frac{x_1 + x_2}{2}, \frac{y_1 + y_2}{2}\right) \rightarrow (x_1, y_1) = (4, 1)$ and $(x_2, y_2) = (2, 3) \rightarrow M = \left(\frac{4+2}{2}, \frac{1+3}{2}\right) \rightarrow$
$\left(\frac{6}{2}, \frac{4}{2}\right) \rightarrow M(3, 2)$

17) $(x_1, y_1) = (3, 6)$ and $(x_2, y_2) = (5, 4) \rightarrow M = \left(\frac{3+5}{2}, \frac{6+4}{2}\right) \rightarrow \left(\frac{8}{2}, \frac{10}{2}\right) \rightarrow M(4, 5)$

18) $(x_1, y_1) = (7, 1)$ and $(x_2, y_2) = (1, 3) \rightarrow M = \left(\frac{7+1}{2}, \frac{1+3}{2}\right) \rightarrow \left(\frac{8}{2}, \frac{4}{2}\right) \rightarrow M(4, 2)$

19) $(x_1, y_1) = (2, 8)$ and $(x_2, y_2) = (2, 10) \rightarrow M = \left(\frac{2+2}{2}, \frac{8+10}{2}\right) \rightarrow \left(\frac{4}{2}, \frac{18}{2}\right) \rightarrow M(2, 9)$

20) $(x_1, y_1) = (3, -2)$ and $(x_2, y_2) = (-1, 6) \rightarrow M = \left(\frac{3-1}{2}, \frac{-2+6}{2}\right) \rightarrow \left(\frac{2}{2}, \frac{4}{2}\right) \rightarrow M(1, 2)$

21) $(x_1, y_1) = (-1, -3)$ and $(x_2, y_2) = (1, 5) \rightarrow M = \left(\frac{-1+1}{2}, \frac{-3+5}{2}\right) \rightarrow \left(\frac{0}{2}, \frac{2}{2}\right) \rightarrow M(0, 1)$

22) $(x_1, y_1) = (1, 4)$ and $(x_2, y_2) = (-7, 6) \rightarrow M = \left(\frac{1-7}{2}, \frac{4+6}{2}\right) \rightarrow \left(\frac{-6}{2}, \frac{10}{2}\right) \rightarrow M(-3, 5)$

23) $(x_1, y_1) = (-3, 5)$ and $(x_2, y_2) = (7, -9) \rightarrow M = \left(\frac{-3+7}{2}, \frac{5-9}{2}\right) \rightarrow \left(\frac{4}{2}, \frac{-4}{2}\right) \rightarrow M(2, -2)$

24) $(x_1, y_1) = (-8, -1)$ and $(x_2, y_2) = (-4, 2) \rightarrow d = \sqrt{(x_2 - x_1)^2 + (y_2 - y_1)^2} =$
$\sqrt{(-4 - (-8))^2 + (2 - (-1))^2} = \sqrt{(4)^2 + (3)^2} = \sqrt{16 + 9} = \sqrt{25} = 5$

25) $(x_1, y_1) = (-15, 2)$ and $(x_2, y_2) = (5, -13) \rightarrow d = \sqrt{(x_2 - x_1)^2 + (y_2 - y_1)^2} =$
$\sqrt{(5 - (-15))^2 + (-13 - 2)^2} = \sqrt{(20)^2 + (-15)^2} = \sqrt{400 + 225} = \sqrt{625} = 25$

26) $(x_1, y_1) = (-1, 11)$ and $(x_2, y_2) = (-7, 3) \rightarrow d = \sqrt{(x_2 - x_1)^2 + (y_2 - y_1)^2} =$
$\sqrt{(-7 - (-1))^2 + (3 - 11)^2} = \sqrt{(-6)^2 + (-8)^2} = \sqrt{36 + 64} = \sqrt{100} = 10$

27) $(x_1, y_1) = (0, 11)$ and $(x_2, y_2) = (9, 11) \rightarrow d = \sqrt{(x_2 - x_1)^2 + (y_2 - y_1)^2} =$
$\sqrt{(9 - 0)^2 + (11 - 11)^2} = \sqrt{(9)^2 + (0)^2} = \sqrt{81} = 9$

28) $(x_1, y_1) = (-2, 4)$ and $(x_2, y_2) = (3, -8) \rightarrow d = \sqrt{(x_2 - x_1)^2 + (y_2 - y_1)^2} =$
$\sqrt{(3 - (-2))^2 + (-8 - 4)^2} = \sqrt{(5)^2 + (-12)^2} = \sqrt{25 + 144} = \sqrt{169} = 13$

29) $(x_1, y_1) = (6, -7)$ and $(x_2, y_2) = (-9, 1) \rightarrow d = \sqrt{(x_2 - x_1)^2 + (y_2 - y_1)^2} =$
$\sqrt{(-9 - 6)^2 + (1 - (-7))^2} = \sqrt{(-15)^2 + (8)^2} = \sqrt{225 + 64} = \sqrt{289} = 17$

30) $(x_1, y_1) = (8, -4)$ and $(x_2, y_2) = (-4, -20) \rightarrow d = \sqrt{(x_2 - x_1)^2 + (y_2 - y_1)^2} =$
$\sqrt{(-4 - 8)^2 + (-20 - (-4))^2} = \sqrt{(-12)^2 + (-16)^2} = \sqrt{144 + 256} = \sqrt{400} = 20$

31) $(x_1, y_1) = (5, 1)$ and $(x_2, y_2) = (9, -2) \rightarrow d = \sqrt{(x_2 - x_1)^2 + (y_2 - y_1)^2} =$
$\sqrt{(9 - 5)^2 + (-2 - 1)^2} = \sqrt{(4)^2 + (-3)^2} = \sqrt{16 + 9} = \sqrt{25} = 5$

32) $(x_1, y_1) = (-8, -17)$ and $(x_2, y_2) = (2, 7) \rightarrow d = \sqrt{(x_2 - x_1)^2 + (y_2 - y_1)^2} =$
$\sqrt{(2 - (-8))^2 + (7 - (-17))^2} = \sqrt{(10)^2 + (-24)^2} = \sqrt{100 + 576} = \sqrt{676} = 26$

33) $(x_1, y_1) = (18, 21)$ and $(x_2, y_2) = (-12, 5) \rightarrow d = \sqrt{(x_2 - x_1)^2 + (y_2 - y_1)^2} =$
$\sqrt{(-12 - 18)^2 + (5 - 21)^2} = \sqrt{(-30)^2 + (-16)^2} = \sqrt{900 + 256} = \sqrt{1,156} = 34$

 **Polynomials**

Math topics that you'll learn in this chapter:

1. Simplifying Polynomials
2. Adding and Subtracting Polynomials
3. Multiplying Monomials
4. Multiplying and Dividing Monomials
5. Multiplying a Polynomial and a Monomial
6. Multiplying Binomials
7. Factoring Trinomials

91

Simplifying Polynomials

☆ To simplify Polynomials, find "like" terms. (they have same variables with same power).

☆ Use "FOIL". (First–Out–In–Last) for binomials:

$$(x + a)(x + b) = x^2 + (b + a)x + ab$$

☆ Add or Subtract "like" terms using order of operation.

Examples:

Example 1. Simplify this expression. $2x(3x - 4) - 6x =$

Solution: Use Distributive Property: $2x(3x - 4) = 6x^2 - 8x$

Now, combine like terms: $2x(3x - 4) - 6x = 6x^2 - 8x - 6x = 6x^2 - 14x$

Example 2. Simplify this expression. $(x + 5)(x + 7) =$

Solution: First, apply the FOIL method: $(a + b)(c + d) = ac + ad + bc + bd$

$(x + 5)(x + 7) = x^2 + 5x + 7x + 35$

Now combine like terms: $x^2 + 5x + 7x + 35 = x^2 + 12x + 35$

Example 3. Simplify this expression. $3x(-4x + 5) + 2x^2 - 5x =$

Solution: Use Distributive Property: $3x(-4x + 5) = -12x^2 + 15x$

Then: $3x(-4x + 5) + 2x^2 - 5x = -12x^2 + 15x + 2x^2 - 5x$

Now combine like terms: $-12x^2 + 2x^2 = -10x^2$, and $15x - 5x = 10x$

The simplified form of the expression: $-12x^2 + 15x + 2x^2 - 5x = -10x^2 + 10x$

Adding and Subtracting Polynomials

☆ Adding polynomials is just a matter of combining like terms, with some order of operations considerations thrown in.

☆ Be careful with the minus signs, and don't confuse addition and multiplication!

☆ For subtracting polynomials, sometimes you need to use the Distributive Property: $a(b + c) = ab + ac$, $a(b - c) = ab - ac$

Examples:

Example 1. Simplify the expressions. $(-5x^2 + 2x^3) - (-4x^3 + 2x^2) =$

Solution: First, use Distributive Property:

$-(-4x^3 + 2x^2) = 4x^3 - 2x^2$

$\rightarrow (-5x^2 + 2x^3) - (-4x^3 + 2x^2) = -5x^2 + 2x^3 + 4x^3 - 2x^2$

Now combine like terms: $2x^3 + 4x^3 = 6x^3$ and $-5x^2 - 2x^2 = -7x^2$

Then: $(x^2 - 2x^3) - (x^3 - 3x^2) = 6x^3 - 7x^2$

Example 2. Add expressions. $(2x^3 + 8) + (-x^3 + 4x^2) =$

Solution: Remove parentheses:

$$(2x^3 + 8) + (-x^3 + 4x^2) = 2x^3 + 8 - x^3 + 4x^2$$

Now combine like terms: $2x^3 + 8 - x^3 + 4x^2 = x^3 + 4x^2 + 8$

Example 3. Simplify the expressions. $(x^2 + 7x^3) - (11x^2 - 4x^3) =$

Solution: First, use Distributive Property: $-(11x^2 - 4x^3) = -11x^2 + 4x^3 \rightarrow$

$$(x^2 + 7x^3) - (11x^2 - 4x^3) = x^2 + 7x^3 - 11x^2 + 4x^3$$

Now combine like terms and write in standard form:

$x^2 + 7x^3 - 11x^2 + 4x^3 = 11x^3 - 10x^2$

Multiplying Monomials

✪ A monomial is a polynomial with just one term: Examples: $2x$ or $7y^2$.

✪ When you multiply monomials, first multiply the coefficients (a number placed before and multiplying the variable) and then multiply the variables using multiplication property of exponents.

$$x^a \times x^b = x^{a+b}$$

Examples:

Example 1. Multiply expressions. $3x^4y^5 \times 6x^2y^3$

Solution: Find the same variables and use multiplication property of exponents:
$x^a \times x^b = x^{a+b}$
$x^4 \times x^2 = x^{4+2} = x^6$ and $y^5 \times y^3 = y^{5+3} = y^8$
Then, multiply coefficients and variables: $3x^4y^5 \times 6x^2y^3 = 18x^6y^8$

Example 2. Multiply expressions. $5a^5b^9 \times 3a^2b^8 =$

Solution: Use the multiplication property of exponents: $x^a \times x^b = x^{a+b}$
$a^5 \times a^2 = a^{5+2} = a^7$ and $b^9 \times b^8 = b^{9+8} = b^{17}$
Then: $5a^5b^9 \times 3a^2b^8 = 15a^7b^{17}$

Example 3. Multiply. $6x^3y^2z^4 \times 2x^2y^8z^6$

Solution: Use the multiplication property of exponents: $x^a \times x^b = x^{a+b}$
$x^3 \times x^2 = x^{3+2} = x^5$, $y^2 \times y^8 = y^{2+8} = y^{10}$ and $z^4 \times z^6 = z^{4+6} = z^{10}$
Then: $6x^3y^2z^4 \times 2x^2y^8z^6 = 12x^5y^{10}z^{10}$

Example 4. Simplify. $(2a^3b^6)(-5a^7b^{12}) =$

Solution: Use the multiplication property of exponents: $x^a \times x^b = x^{a+b}$
$a^3 \times a^7 = a^{3+7} = a^{10}$ and $b^6 \times b^{12} = b^{6+12} = b^{18}$
Then: $(2a^3b^6)(-5a^7b^{12}) = -10a^{10}b^{18}$

Multiplying and Dividing Monomials

☆ When you divide or multiply two monomials, you need to divide or multiply their coefficients and then divide or multiply their variables.

☆ In case of exponents with the same base, for Division, subtract their powers, for Multiplication, add their powers.

☆ Exponent's Multiplication and Division rules:

$$x^a \times x^b = x^{a+b} , \qquad \frac{x^a}{x^b} = x^{a-b}$$

Examples:

Example 1. Multiply expressions. $(5x^4)(3x^9) =$

Solution: Use multiplication property of exponents:
$x^a \times x^b = x^{a+b} \rightarrow x^4 \times x^9 = x^{13}$
Then: $(5x^4)(3x^9) = 15x^{13}$

Example 2. Divide expressions. $\frac{18x^3y^6}{9x^2y^4} =$

Solution: Use division property of exponents:
$\frac{x^a}{x^b} = x^{a-b} \rightarrow \frac{x^3}{x^2} = x^{3-2} = x^1$ and $\frac{y^6}{y^4} = y^{6-4} = y^2$
Then: $\frac{18x^3y^6}{9x^2y^4} = 2xy^2$

Example 3. Divide expressions. $\frac{51a^4b^{11}}{3a^2b^5}$

Solution: Use division property of exponents:
$\frac{x^a}{x^b} = x^{a-b} \rightarrow \frac{a^4}{a^2} = a^{4-2} = a^2$ and $\frac{b^{11}}{b^5} = b^{11-5} = b^6$
Then. $\frac{51a^4b^{11}}{3a^2b^5} = 17a^2b^6$

Multiplying a Polynomial and a Monomial

✰ When multiplying monomials, use the product rule for exponents.

$$x^a \times x^b = x^{a+b}$$

✰ When multiplying a monomial by a polynomial, use the distributive property.

$$a \times (b + c) = a \times b + a \times c = ab + ac$$
$$a \times (b - c) = a \times b - a \times c = ab - ac$$

Examples:

Example 1. Multiply expressions. $5x(4x + 7)$

Solution: Use Distributive Property:

$5x(4x + 7) = (5x \times 4x) + (5x \times 7) = 20x^2 + 35x$

Example 2. Multiply expressions. $y(2x^2 + 3y^2)$

Solution: Use Distributive Property:

$y(2x^2 + 3y^2) = y \times 2x^2 + y \times 3y^2 = 2x^2y + 3y^3$

Example 3. Multiply. $-2x(-x^2 + 3x + 6)$

Solution: Use Distributive Property:

$-2x(-x^2 + 3x + 6) = (-2x)(-x^2) + (-2x) \times (3x) + (-2x) \times (6) =$

Now simplify:

$(-2x)(-x^2) + (-2x) \times (3x) + (-2x) \times (6) = 2x^3 - 6x^2 - 12x$

Multiplying Binomials

☆ A binomial is a polynomial that is the sum or the difference of two terms, each of which is a monomial.

☆ To multiply two binomials, use the "FOIL" method. (First–Out–In–Last)

$$(x + a)(x + b) = x \times x + x \times b + a \times x + a \times b = x^2 + bx + ax + ab$$

Examples:

Example 1. Multiply Binomials. $(x - 5)(x + 4) =$

Solution: Use "FOIL". (First–Out–In–Last):

$(x - 5)(x + 4) = x^2 - 5x + 4x - 20$

Then combine like terms: $x^2 - 5x + 4x - 20 = x^2 - x - 20$

Example 2. Multiply. $(x + 3)(x + 6) =$

Solution: Use "FOIL". (First–Out–In–Last):

$(x + 3)(x + 6) = x^2 + 3x + 6x + 18$

Then simplify: $x^2 + 3x + 6x + 18 = x^2 + 9x + 18$

Example 3. Multiply. $(x - 8)(x + 4) =$

Solution: Use "FOIL". (First–Out–In–Last):

$(x - 8)(x + 4) = x^2 - 8x + 4x - 32$

Then simplify: $x^2 - 8x + 4x - 32 = x^2 - 4x - 32$

Example 4. Multiply Binomials. $(x - 6)(x - 3) =$

Solution: Use "FOIL". (First–Out–In–Last):

$(x - 6)(x - 3) = x^2 - 6x - 3x + 18$

Then combine like terms: $x^2 - 6x - 3x + 18 = x^2 - 9x + 18$

Factoring Trinomials

To factor trinomials, you can use following methods:

☆ "FOIL": $(x + a)(x + b) = x^2 + (b + a)x + ab$

☆ "Difference of Squares":

$$a^2 - b^2 = (a + b)(a - b)$$
$$a^2 + 2ab + b^2 = (a + b)(a + b)$$
$$a^2 - 2ab + b^2 = (a - b)(a - b)$$

☆ "Reverse FOIL": $x^2 + (b + a)x + ab = (x + a)(x + b)$

Examples:

Example 1. Factor this trinomial. $x^2 - 3x - 18$

Solution: Break the expression into groups. You need to find two numbers that their product is -18 and their sum is -3. (remember "Reverse FOIL": $x^2 + (b + a)x + ab = (x + a)(x + b)$). Those two numbers are 3 and -6. Then:
$$x^2 - 3x - 18 = (x^2 + 3x) + (-6x - 18)$$
Now factor out x from $x^2 + 3x : x(x + 3)$, and factor out -6 from
$-6x - 18: -6(x + 3)$; Then: $(x^2 + 3x) + (-6x - 18) = x(x + 3) - 6(x + 3)$
Now factor out like term: $(x + 3)$. Then: $(x + 3)(x - 6)$

Example 2. Factor this trinomial. $2x^2 - 4x - 48$

Solution: Break the expression into groups: $(2x^2 + 8x) + (-12x - 48)$
Now factor out $2x$ from $2x^2 + 8x : 2x(x + 4)$, and factor out -12 from
$- 12x - 48: -12(x + 4)$; Then: $2x(x + 4) - 12(x + 4)$, now factor out like term:
$(x + 4) \rightarrow 2x(x + 4) - 12(x + 4) = (x + 4)(2x - 12)$

Day 8: Practices

✎ Simplify each polynomial.

1) $4(3x + 2) =$

2) $7(6x - 3) =$

3) $x(4x + 5) + 6x =$

4) $2x(x - 4) + 8x =$

5) $3(5x + 3) - 9x =$

6) $x(6x - 5) - 4x^2 + 11 =$

7) $-x^2 + 7 + 3x(x + 2) =$

8) $6x^2 - 7 + 3x(5x - 7) =$

✎ Add or subtract polynomials.

9) $(x^2 + 5) + (3x^2 - 2) =$

10) $(4x^2 - 5x) - (x^2 + 7x) =$

11) $(6x^3 - 2x^2) + (3x^3 - 7x^2) =$

12) $(6x^3 - 7x) - (9x^3 - 3x) =$

13) $(9x^3 + 5x^2) + (12x^2 - 7) =$

14) $(5x^3 - 8) - (2x^3 - 6x^2) =$

15) $(10x^3 + 4x) - (7x^3 - 5x) =$

16) $(12x^3 - 7x) - (3x^3 + 9x) =$

✎ Find the products. (Multiplying Monomials)

17) $5x^3 \times 6x^5 =$

18) $3x^4 \times 4x^3 =$

19) $-7a^3b \times 3a^2b^5 =$

20) $-5x^2y^3z \times 6x^6y^4z^5 =$

21) $-2a^3bc \times (-4a^8b^7) =$

22) $9u^6t^5 \times (-2u^2t) =$

23) $14x^2z \times 2x^6y^8z =$

24) $-12x^7y^6z \times 3xy^8 =$

25) $-9a^2b^3c \times 3a^7b^6 =$

26) $-11x^9y^7 \times (-6x^4y^3) =$

✍ **Simplify each expression. (Multiplying and Dividing Monomials)**

27) $(4x^3y^4)(2x^4y^3) =$

28) $(7x^2y^5)(3x^3y^6) =$

29) $(5x^9y^6)(8x^6y^9) =$

30) $(13a^4b^7)(2a^6b^9) =$

31) $\frac{54x^6y^3}{9x^4y} =$

32) $\frac{28x^5y^7}{4x^3y^4} =$

33) $\frac{32x^{17}y^{12}}{8x^{13}y^9} =$

34) $\frac{40x^7y^{19}}{5x^2y^{14}} =$

✍ **Find each product. (Multiplying a Polynomial and a Monomial)**

35) $6(4x - 2y) =$

36) $4x(5x + y) =$

37) $8x(x - 2y) =$

38) $x(3x^2 + 4x - 6) =$

39) $4x(2x^2 + 7x + 4) =$

40) $8x(3x^2 - 7x - 3) =$

✍ **Find each product. (Multiplying Binomials)**

41) $(x - 4)(x + 5) =$

42) $(x - 3)(x + 3) =$

43) $(x + 8)(x + 7) =$

44) $(x - 5)(x + 9) =$

45) $(2x + 4)(x - 6) =$

46) $(2x - 11)(x + 6) =$

✍ **Factor each trinomial.**

47) $x^2 + 2x - 15 =$

48) $x^2 - x - 42 =$

49) $x^2 - 14x + 49 =$

50) $x^2 - 7x - 60 =$

51) $2x^2 + 6x - 20 =$

52) $3x^2 + 13x - 10 =$

Day 8: Answers

1) $4(3x + 2) = (4 \times 3x) + (4 \times 2) = 12x + 8$

2) $7(6x - 3) = (7 \times 6x) - (7 \times 3) = 42x - 21$

3) $x(4x + 5) + 6x = (x \times 4x) + (x \times 5) + 6x = 4x^2 + 5x + 6x = 4x^2 + 11x$

4) $2x(x - 4) + 8x = (2x \times x) + (2x \times (-4)) + 8x = 2x^2 - 8x + 8x = 2x^2$

5) $3(5x + 3) - 9x = (3 \times 5x) + (3 \times 3) - 9x = 15x - 9x + 9 = 6x + 9$

6) $x(6x - 5) - 4x^2 + 11 = (x \times 6x) + (x \times (-5)) - 4x^2 + 11 =$
$6x^2 - 4x^2 - 5x + 11 = 2x^2 - 5x + 11$

7) $-x^2 + 7 + 3x(x + 2) = -x^2 + 7 + (3x \times x) + (3x \times 2) = -x^2 + 7 + 3x^2 + 6x =$
$2x^2 + 6x + 7$

8) $6x^2 - 7 + 3x(5x - 7) = 6x^2 - 7 + (3x \times 5x) + (3x \times (-7)) =$
$6x^2 - 7 + 15x^2 - 21x = 21x^2 - 21x - 7$

9) $(x^2 + 5) + (3x^2 - 2) = x^2 + 3x^2 + 5 - 2 = 4x^2 + 3$

10) $(4x^2 - 5x) - (x^2 + 7x) = 4x^2 - x^2 - 5x - 7x = 3x^2 - 12x$

11) $(6x^3 - 2x^2) + (3x^3 - 7x^2) = 6x^3 + 3x^3 - 2x^2 - 7x^2 = 9x^3 - 9x^2$

12) $(6x^3 - 7x) - (9x^3 - 3x) = 6x^3 - 9x^3 - 7x + 3x = -3x^3 - 4x$

13) $(9x^3 + 5x^2) + (12x^2 - 7) = 9x^3 + 5x^2 + 12x^2 - 7 = 9x^3 + 17x^2 - 7$

14) $(5x^3 - 8) - (2x^3 - 6x^2) = 5x^3 - 2x^3 + 6x^2 - 8 = 3x^3 + 6x^2 - 8$

15) $(10x^3 + 4x) - (7x^3 - 5x) = 10x^3 - 7x^3 + 4x + 5x = 3x^3 + 9x$

16) $(12x^3 - 7x) - (3x^3 + 9x) = 12x^3 - 3x^3 - 7x - 9x = 9x^3 - 16x$

17) $5x^3 \times 6x^5 \rightarrow 5 \times 6 = 30, \; x^3 \times x^5 = x^{3+5} = x^8 \rightarrow 5x^3 \times 6x^5 = 30x^8$

18) $3x^4 \times 4x^3 \rightarrow 3 \times 4 = 12, \; x^4 \times x^3 = x^{4+3} = x^7 \rightarrow 3x^4 \times 4x^3 = 12x^7$

19) $-7a^3b \times 3a^2b^5 \rightarrow -7 \times 3 = -21, \; a^3 \times a^2 = a^{3+2} = a^5, \; b \times b^5 = b^{1+5} = b^6 \rightarrow$
$-7a^3b \times 3a^2b^5 = -21a^5b^6$

20) $-5x^2y^3z \times 6x^6y^4z^5 \rightarrow -5 \times 6 = -30, \; x^2 \times x^6 = x^{2+6} = x^8, y^3 \times y^4 = y^{3+4} = y^7,$
$z \times z^5 = z^{1+5} = z^6 \rightarrow -5x^2y^3z \times 6x^6y^4z^5 = -30x^8y^7z^6$

21) $-2a^3bc \times (-4a^8b^7) \rightarrow -2 \times (-4) = 8,\ a^3 \times a^8 = a^{3+8} = a^{11}, b \times b^7 = b^{1+7} = b^8 \rightarrow -2a^3bc \times (-4a^8b^7) = 8a^{11}b^8c$

22) $9u^6t^5 \times (-2u^2t) \rightarrow 9 \times (-2) = -18,\ u^6 \times u^2 = u^{6+2} = u^8, t^5 \times t^1 = t^{5+1} = t^6 \rightarrow 9u^6t^5 \times (-2u^2t) = -18u^8t^6$

23) $14x^2z \times 2x^6y^8z \rightarrow 14 \times 2 = 28, x^2 \times x^6 = x^{2+6} = x^8, z \times z = z^{1+1} = z^2 \rightarrow 14x^2z \times 2x^6y^8z = 28x^8y^8z^2$

24) $-12x^7y^6z \times 3xy^8 \rightarrow -12 \times 3 = -36, x^7 \times x = x^{1+7} = x^8,\ y^6 \times y^8 = y^{6+8} = y^{14} \rightarrow -12x^7y^6z \times 3xy^8 = -36x^8y^{14}z$

25) $-9a^2b^3c \times 3a^7b^6 \rightarrow -9 \times 3 = -27, a^2 \times a^7 = a^{2+7} = a^9,\ b^3 \times b^6 = b^{3+6} = b^9 \rightarrow -9a^2b^3c \times 3a^7b^6 = -27a^9b^9c$

26) $-11x^9y^7 \times (-6x^4y^3) \rightarrow -11 \times (-6) = 66, x^9 \times x^4 = x^{9+4} = x^{13}, y^7 \times y^3 = y^{7+3} = y^{10} \rightarrow -11x^9y^7 \times (-6x^4y^3) = 66x^{13}y^{10}$

27) $(4x^3y^4)(2x^4y^3) \rightarrow 4 \times 2 = 8, x^3 \times x^4 = x^{3+4} = x^7,\ y^4 \times y^3 = y^{4+3} = y^7 \rightarrow (4x^3y^4)(2x^4y^3) = 8x^7y^7$

28) $(7x^2y^5)(3x^3y^6) \rightarrow 7 \times 3 = 21, x^2 \times x^3 = x^{2+3} = x^5,\ y^5 \times y^6 = y^{5+6} = y^{11} \rightarrow (7x^2y^5)(3x^3y^6) = 21x^5y^{11}$

29) $(5x^9y^6)(8x^6y^9) \rightarrow 5 \times 8 = 40, x^9 \times x^6 = x^{9+6} = x^{15},\ y^6 \times y^9 = y^{6+9} = y^{15} \rightarrow (5x^9y^6)(8x^6y^9) = 40x^{15}y^{15}$

30) $(13a^4b^7)(2a^6b^9) \rightarrow 13 \times 2 = 26, a^4 \times a^6 = a^{4+6} = a^{10},\ b^7 \times b^9 = b^{7+9} = b^{16} \rightarrow (13a^4b^7)(2a^6b^9) = 26a^{10}b^{16}$

31) $\frac{54x^6y^3}{9x^4y} \rightarrow \frac{54}{9} = 6,\ \frac{x^6}{x^4} = x^{6-4} = x^2,\ \frac{y^3}{y} = y^{3-1} = y^2 \rightarrow \frac{54x^6y^3}{9x^4y} = 6x^2y^2$

32) $\frac{28x^5y^7}{4x^3y^4} \rightarrow \frac{28}{4} = 7,\ \frac{x^5}{x^3} = x^{5-3} = x^2,\ \frac{y^7}{y^4} = y^{7-4} = y^3 \rightarrow \frac{28x^5y^7}{4x^3y^4} = 7x^2y^3$

33) $\frac{32x^{17}y^{12}}{8x^{13}y^9} \rightarrow \frac{32}{8} = 4,\ \frac{x^{17}}{x^{13}} = x^{17-13} = x^4,\ \frac{y^{12}}{y^9} = y^{12-9} = y^3 \rightarrow \frac{32x^{17}y^{12}}{8x^{13}y^9} = 4x^4y^3$

34) $\frac{40x^7y^{19}}{5x^2y^{14}} = \rightarrow \frac{40}{5} = 8,\ \frac{x^7}{x^2} = x^{7-2} = x^5,\ \frac{y^{19}}{y^{14}} = y^{19-14} = y^5 \rightarrow \frac{40x^7y^{19}}{5x^2y^{14}} = 8x^5y^5$

35) $6(4x - 2y) = (6 \times 4x) - (6 \times 2y) = 24x - 12y$

36) $4x(5x + y) = (4x \times 5x) + (4x \times y) = 20x^2 + 4xy$

37) $8x(x - 2y) = (8x \times x) - (8x \times 2y) = 8x^2 - 16xy$

38) $x(3x^2 + 4x - 6) = (x \times 3x^2) + (x \times 4x) + (x \times (-6)) = 3x^3 + 4x^2 - 6x$

39) $4x(2x^2 + 7x + 4) = (4x \times 2x^2) + (4x \times 7x) + (4x \times 4) = 8x^3 + 28x^2 + 16x$

40) $8x(3x^2 - 7x - 3) = (8x \times 3x^2) + (8x \times (-7x)) + (8x \times (-3)) = 24x^3 - 56x^2 - 24x$

41) $(x - 4)(x + 5) = (x \times x) + (x \times 5) + (-4 \times x) + (-4 \times 5) = x^2 + 5x - 4x - 20 = $
$x^2 + x - 20$

42) $(x - 3)(x + 3) = (x \times x) + (x \times 3) + (-3 \times x) + (-3 \times 3) = x^2 + 3x - 3x - 9 = x^2 - 9$

43) $(x + 8)(x + 7) = (x \times x) + (x \times 7) + (8 \times x) + (8 \times 7) = x^2 + 7x + 8x + 56 = $
$x^2 + 15x + 56$

44) $(x - 5)(x + 9) = (x \times x) + (x \times 9) + (-5 \times x) + (-5 \times 9) = x^2 + 9x - 5x - 45 = $
$x^2 + 4x - 45$

45) $(2x + 4)(x - 6) = (2x \times x) + (2x \times (-6)) + (4 \times x) + (4 \times (-6)) = $
$2x^2 - 12x + 4x - 24 = 2x^2 - 8x - 24$

46) $(2x - 11)(x + 6) = (2x \times x) + (2x \times 6) + ((-11) \times x) + ((-11) \times 6) = $
$2x^2 + 12x - 11x - 66 = 2x^2 + x - 66$

47) $x^2 + 2x - 15 \rightarrow$ (Use this rule: $x^2 + (b + a)x + ab = (x + a)(x + b)$). Then:
$x^2 + 2x - 15 = x^2 + (5 - 3)x + (5 \times (-3)) = (x + 5)(x - 3)$

48) $x^2 - x - 42 = x^2 + (-7 + 6)x + ((-7) \times 6) = (x - 7)(x + 6)$

49) $x^2 - 14x + 49 = x^2 + (-7 - 7)x + ((-7) \times (-7)) = (x - 7)(x - 7)$

50) $x^2 - 7x - 60x^2 = x^2 + (-12 + 5)x + ((-12) \times 5) = (x - 12)(x + 5)$

51) $2x^2 + 6x - 20 = 2x^2 + (10x - 4x) - 20 = (2x^2 + 10x) + (-4x - 20) = $
$2x(x + 5) - 4(x + 5) = (2x - 4)(x + 5)$

52) $3x^2 + 13x - 10 = (3x^2 + 15x) + (-2x - 10) = 3x(x + 5) - 2(x + 5) = $
$(3x - 2)(x + 5)$

Geometry and Solid Figures

Math topics that you'll learn in this chapter:

1. The Pythagorean Theorem
2. Complementary and Supplementary angles
3. Parallel lines and Transversals
4. Triangles
5. Special Right Triangles
6. Polygons
7. Circles
8. Trapezoids
9. Cubes
10. Rectangle Prisms
11. Cylinder

The Pythagorean Theorem

☆ You can use the Pythagorean Theorem to find a missing side in a right triangle.

☆ In any right triangle: $a^2 + b^2 = c^2$

Examples:

Example 1. Right triangle ABC (not shown) has two legs of lengths 3 cm (AB) and 4 cm (AC). What is the length of the hypotenuse of the triangle (side BC)?

Solution: Use Pythagorean Theorem: $a^2 + b^2 = c^2$, $a = 3$ and $b = 4$

Then: $a^2 + b^2 = c^2 \rightarrow 3^2 + 4^2 = c^2 \rightarrow 9 + 16 = c^2 \rightarrow 25 = c^2 \rightarrow c = \sqrt{25} = 5$

The length of the hypotenuse is 5 cm.

Example 2. Find the hypotenuse of this triangle.

Solution: Use Pythagorean Theorem: $a^2 + b^2 = c^2$

Then: $a^2 + b^2 = c^2 \rightarrow 15^2 + 8^2 = c^2 \rightarrow 225 + 64 = c^2$

$c^2 = 289 \rightarrow c = \sqrt{289} = 17$

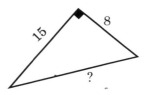

Example 3. Find the length of the missing side in this triangle.

Solution: Use Pythagorean Theorem: $a^2 + b^2 = c^2$

Then: $a^2 + b^2 = c^2 \rightarrow 20^2 + b^2 = 25^2 \rightarrow 400 + b^2 = 625 \rightarrow$

$b^2 = 625 - 400 \rightarrow b^2 = 225 \rightarrow b = \sqrt{225} = 15$

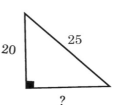

Complementary and Supplementary angles

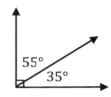

☆ Two angles with a sum of 90 degrees are called complementary angles.

☆ Two angles with a sum of 180 degrees are Supplementary angles.

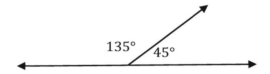

Examples:

Example 1. Find the missing angle.

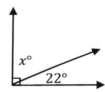

Solution: Notice that the two angles form a right angle. This means that the angles are complementary, and their sum is 90.
Then: $22° + x = 90° \rightarrow x = 90° - 22° = 68°$
The missing angle is 68 degrees. $x = 68°$

Example 2. Angles Q and S are supplementary. What is the measure of angle Q if angle S is 45 degrees?

Solution: Q and S are supplementary $\rightarrow Q + S = 180 \rightarrow Q + 45 = 180 \rightarrow$
$$Q = 180 - 45 = 135°$$

Example 3. Angles x and y are complementary. What is the measure of angle x if angle y is 27 degrees?

Solution: Angles x and y are complementary $\rightarrow x + y = 90 \rightarrow x + 27 = 90 \rightarrow$
$$x = 90 - 27 = 63°$$

Parallel lines and Transversals

✭ When a line (transversal) intersects two parallel lines in the same plane, eight angles are formed. In the following diagram, a transversal intersects two parallel lines. Angles $1, 3, 5$ and 7 are congruent. Angles $2, 4, 6,$ and 8 are also congruent.

✭ In the following diagram, the following angles are supplementary angles (their sum is 180):

❖ Angles 1 and 8

❖ Angles 2 and 7

❖ Angles 3 and 6

❖ Angles 4 and 5

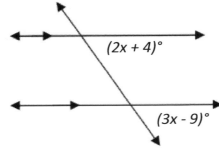

Example:

In the following diagram, two parallel lines are cut by a transversal. What is the value of x?

Solution: The two angles $2x + 4$ and $3x - 9$ are equivalent.

That is: $2x + 4 = 3x - 9$

Now, solve for x:

$2x + 4 + 9 = 3x - 9 + 9$

$\rightarrow 2x + 13 = 3x \rightarrow 2x + 13 - 2x = 3x - 2x \rightarrow$

$13 = x$

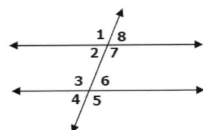

Triangles

★ In any triangle, the sum of all angles is 180 degrees.

★ Area of a triangle $= \frac{1}{2}$(base×height)

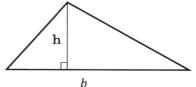

Examples:

Example 1. What is the area of this triangles?

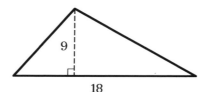

Solution: Use the area formula:

Area$= \frac{1}{2}$(base×height)

base$= 18$ and height$= 9$, Then:

Area$= \frac{1}{2}(18 \times 9) = \frac{162}{2} = 81$

Example 2. What is the area of this triangles?

Solution: Use the area formula:

Area$= \frac{1}{2}$(base×height)

base$= 18$ and height$= 7$; Area$= \frac{1}{2}(18 \times 7) = \frac{126}{2} = 63$

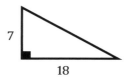

Example 3. What is the missing angle in this triangle?

Solution: In any triangle, the sum of all angles is 180 degrees. Let x be the missing angle.
Then: $63 + 84 + x = 180 \rightarrow 147 + x = 180 \rightarrow$
$x = 180 - 147 = 33º$
The missing angle is 33 degrees.

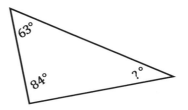

Special Right Triangles

☆ A special right triangle is a triangle whose sides are in a particular ratio. Two special right triangles are $45° - 45° - 90°$ and $30° - 60° - 90°$ triangles.

☆ In a special $45° - 45° - 90°$ triangle, the three angles are $45°$, $45°$ and $90°$. The lengths of the sides of this triangle are in the ratio of $1 : 1 : \sqrt{2}$.

☆ In a special triangle $30° - 60° - 90°$, the three angles are $30° - 60° - 90°$. The lengths of this triangle are in the ratio of $1 : \sqrt{3} : 2$.

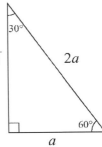

Examples:

Example 1. Find the length of the hypotenuse of a right triangle if the length of the other two sides are both 6 inches.

Solution: This is a right triangle with two equal sides. Therefore, it must be a $45° - 45° - 90°$ triangle. Two equivalent sides are 6 inches. The ratio of sides: $x : x : x\sqrt{2}$
The length of the hypotenuse is $6\sqrt{2}$ inches. $x : x : x\sqrt{2} \rightarrow 6 : 6 : 6\sqrt{2}$

Example 2. The length of the hypotenuse of a right triangle is 6 inches. What are the lengths of the other two sides if one angle of the triangle is $30°$?

Solution: The hypotenuse is 6 inches and the triangle is a $30° - 60° - 90°$ triangle. Then, one side of the triangle is 3 (it's half the side of the hypotenuse) and the other side is $3\sqrt{3}$. (it's the smallest side times $\sqrt{3}$)
$x : x\sqrt{3} : 2x \rightarrow x = 3 \rightarrow x : x\sqrt{3} : 2x = 3 : 3\sqrt{3} : 6$

Polygons

✩ The perimeter of a square $= 4 \times side = 4s$

✩ The perimeter of a rectangle $= 2(width + length)$

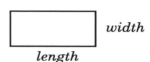

✩ The perimeter of trapezoid $= a + b + c + d$

✩ The perimeter of a regular hexagon $= 6a$

✩ The perimeter of a parallelogram $= 2(l + w)$

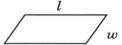

Examples:

Example 1. Find the perimeter of following regular hexagon.

Solution: Since the hexagon is regular, all sides are equal.
Then, the perimeter of the hexagon $= 6 \times (one\ side)$
The perimeter of the hexagon $= 6 \times (one\ side) = 6 \times 7 = 42\ m$

Example 2. Find the perimeter of following trapezoid.

Solution: The perimeter of a trapezoid $= a + b + c + d$
The perimeter of the trapezoid $= 9 + 7 + 13 + 7 = 36\ ft$

Circles

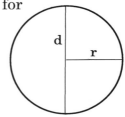

★ In a circle, variable r is usually used for the radius and d for diameter.

★ *Area of a circle* $= \pi r^2$ (π is about 3.14)

★ *Circumference of a circle* $= 2\pi r$

Examples:

Example 1. Find the area of this circle. ($\pi = 3.14$)

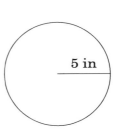

Solution:
Use area formula: $Area = \pi r^2$
$r = 5\ in \rightarrow Area = \pi(5)^2 = 25\pi$, $\pi = 3.14$
Then: $Area = 25 \times 3.14 = 78.5\ in^2$

Example 2. Find the Circumference of this circle. ($\pi = 3.14$)

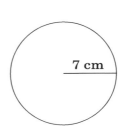

Solution:
Use Circumference formula: $Circumference = 2\pi r$
$r = 7\ cm \rightarrow Circumference = 2\pi(7) = 14\pi$
$\pi = 3.14$, Then: $Circumference = 14 \times 3.14 = 43.96\ cm$

Example 3. Find the area of this circle.

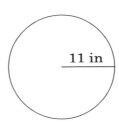

Solution:
Use area formula: $Area = \pi r^2$
$r = 11\ in$, Then: $Area = \pi(11)^2 = 121\pi$, $\pi = 3.14$
$Area = 121 \times 3.14 = 379.94\ in^2$

Trapezoids

☆ A quadrilateral with at least one pair of parallel sides is a trapezoid.

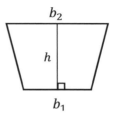

☆ Area of a trapezoid $= \frac{1}{2}h(b_1 + b_2)$

Examples:

Example 1. Calculate the area of this trapezoid.

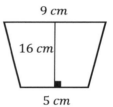

Solution:

Use area formula: $A = \frac{1}{2}h(b_1 + b_2)$

$b_1 = 5 \, cm$, $b_2 = 9 \, cm$ and $h = 16 \, cm$

Then: $A = \frac{1}{2}(16)(9 + 5) = 8(14) = 112 \, cm^2$

Example 2. Calculate the area of this trapezoid.

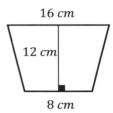

Solution:

Use area formula: $A = \frac{1}{2}h(b_1 + b_2)$

$b_1 = 8 \, cm$, $b_2 = 16 \, cm$ and $h = 12 \, cm$

Then: $A = \frac{1}{2}(12)(8 + 16) = 144 \, cm^2$

Cubes

✩ A cube is a three-dimensional solid object bounded by six square sides.

✩ Volume is the measure of the amount of space inside of a solid figure, like a cube, ball, cylinder or pyramid.

✩ The volume of a cube = $(one\ side)^3$

✩ The surface area of a cube = $6 \times (one\ side)^2$

Examples:

Example 1. Find the volume and surface area of this cube.

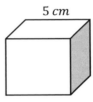

Solution: Use volume formula: $volume = (one\ side)^3$
Then: $volume = (one\ side)^3 = (5)^3 = 125\ cm^3$
Use surface area formula:
$surface\ area\ of\ a\ cube: 6(one\ side)^2 = 6(5)^2 = 6(25) = 150\ cm^2$

Example 2. Find the volume and surface area of this cube.

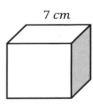

Solution: Use volume formula: $volume = (one\ side)^3$
Then: $volume = (one\ side)^3 = (7)^3 = 343\ cm^3$
Use surface area formula:
$surface\ area\ of\ a\ cube: 6(one\ side)^2 = 6(7)^2 = 6(49) = 294\ cm^2$

Example 3. Find the volume and surface area of this cube.

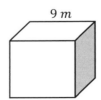

Solution: Use volume formula: $volume = (one\ side)^3$
Then: $volume = (one\ side)^3 = (9)^3 = 729\ m^3$
Use surface area formula:
$surface\ area\ of\ a\ cube: 6(one\ side)^2 = 6(9)^2 = 6(81) = 486\ m^2$

Rectangular Prisms

☆ A rectangular prism is a solid 3-dimensional object with six rectangular faces.

☆ The volume of a Rectangular prism = *Length × Width × Height*

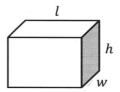

$Volume = l \times w \times h$

$Surface\ area = 2 \times (wh + lw + lh)$

Examples:

Example 1. Find the volume and surface area of this rectangular prism.

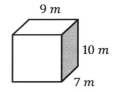

Solution: Use volume formula: $Volume = l \times w \times h$

Then: $Volume = 9 \times 7 \times 10 = 630\ m^3$

Use surface area formula: $Surface\ area = 2 \times (wh + lw + lh)$

Then: $Surface\ area = 2 \times ((7 \times 10) + (9 \times 7) + (9 \times 10))$

$$= 2 \times (70 + 63 + 90) = 2 \times (223) = 446\ m^2$$

Example 2. Find the volume and surface area of this rectangular prism.

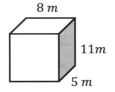

Solution: Use volume formula: $Volume = l \times w \times h$

Then: $Volume = 8 \times 5 \times 11 = 440\ m^3$

Use surface area formula: $Surface\ area = 2 \times (wh + lw + lh)$

Then: $Surface\ area = 2 \times ((5 \times 11) + (8 \times 5) + (8 \times 11))$

$$= 2 \times (55 + 40 + 88) = 2 \times (183) = 366\ m^2$$

Cylinder

☆ A cylinder is a solid geometric figure with straight parallel sides and a circular or oval cross-section.

☆ *Volume of a Cylinder = $\pi(radius)^2 \times height$, $\pi \approx 3.14$*

☆ *Surface area of a cylinder = $2\pi r^2 + 2\pi rh$*

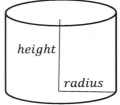

Examples:

Example 1. Find the volume and Surface area of the follow Cylinder.

Solution: Use volume formula:
Volume = $\pi(radius)^2 \times height$
Then: *Volume = $\pi(6)^2 \times 12 = 36\pi \times 12 = 432\pi$*
$\pi = 3.14$, then: *Volume = $432\pi = 432 \times 3.14 = 1,356.48\ cm^3$*
Use surface area formula: *Surface area = $2\pi r^2 + 2\pi rh$*
Then: $2\pi(6)^2 + 2\pi(6)(12) = 2\pi(36) + 2\pi(72) = 72\pi + 144\pi = 216\pi$
$\pi = 3.14$, Then: *Surface area = $216 \times 3.14 = 678.24\ cm^2$*

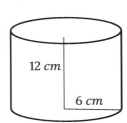

Example 2. Find the volume and Surface area of the follow Cylinder.

Solution: Use volume formula:
Volume = $\pi(radius)^2 \times height$
Then: *Volume = $\pi(2)^2 \times 5 = 4\pi \times 5 = 20\pi$*
$\pi = 3.14$, Then: *Volume = $20\pi = 62.8\ cm^3$*
Use surface area formula: *Surface area = $2\pi r^2 + 2\pi rh$*
Then: $= 2\pi(2)^2 + 2\pi(2)(5) = 2\pi(4) + 2\pi(10) = 8\pi + 20\pi = 28\pi$
$\pi = 3.14$, then: *Surface area = $28 \times 3.14 = 87.92\ cm^2$*

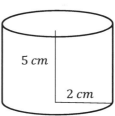

Day 9: Practices

✍ Find the missing side?

1)

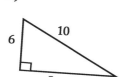

2)

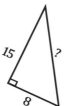

3)

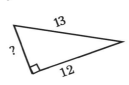

4)

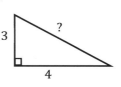

✍ Find the measure of the unknown angle in each triangle.

5)

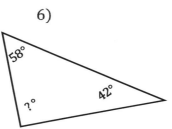

6)

7)

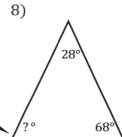

8)

✍ Find the area of each triangle.

9)

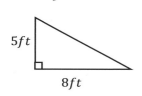

10)

11)

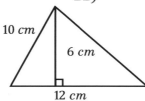

12)

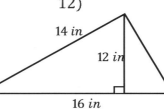

✍ Find the perimeter or circumference of each shape.

13)

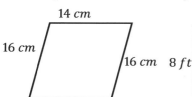

14)

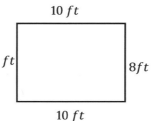

15)

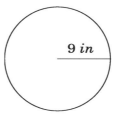

16) regular hexagon

✎ Find the area of each trapezoid.

17)

12 m

8 m

15 m

18)

10 cm

6 cm

14 cm

19)

9 ft

7 ft

13 ft

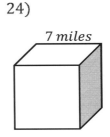

20)

8 cm

6 cm

12 cm

✎ Find the volume of each cube.

21)

5 cm

22)

30 ft

23)

11 in

24)

7 miles

✎ Find the volume of each Rectangular Prism.

25)

9 cm

8 cm

5 cm

26)

14 m

9 m

6 m

27)

12 in

8 in

5 in

✎ Find the volume of each Cylinder. Round your answer to the nearest tenth. ($\pi = 3.14$)

28)

6 cm

14 cm

29)

9 m

12 m

30)

6 cm

10cm

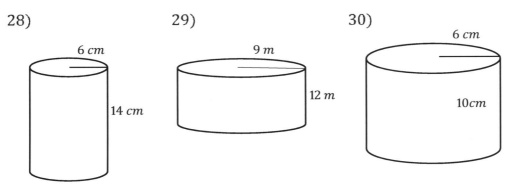

Day 9: Answers

1) Use Pythagorean Theorem: $a^2 + b^2 = c^2$, $a = 6$ and $c = 10$, Then: $6^2 + b^2 = 10^2 \rightarrow 36 + b^2 = 100 \rightarrow b^2 = 100 - 36 = 64 \rightarrow b = \sqrt{64} = 8$

2) Use Pythagorean Theorem: $a^2 + b^2 = c^2$, $a = 15$ and $b = 8$, Then: $15^2 + 8^2 = c^2 \rightarrow 225 + 64 = c^2 \rightarrow c^2 = 289 \rightarrow c = \sqrt{289} = 17$

3) Use Pythagorean Theorem: $a^2 + b^2 = c^2$, $a = 12$ and $c = 13$, Then: $12^2 + b^2 = 13^2 \rightarrow 144 + b^2 = 169 \rightarrow b^2 = 169 - 144 = 25 \rightarrow b = \sqrt{25} = 5$

4) Use Pythagorean Theorem: $a^2 + b^2 = c^2$, $a = 3$ and $b = 4$, Then: $3^2 + 4^2 = c^2 \rightarrow 9 + 16 = c^2 \rightarrow c^2 = 25 \rightarrow c = \sqrt{25} = 5$

5) In any triangle, the sum of all angles is 180 degrees. Then: $85 + 40 + x = 180 \rightarrow 125 + x = 180 \rightarrow x = 180 - 125 = 55$

6) $58 + 42 + x = 180 \rightarrow 100 + x = 180 \rightarrow x = 180 - 100 = 80$

7) $71 + 18 + x = 180 \rightarrow 89 + x = 180 \rightarrow x = 180 - 89 = 91$

8) $28 + 68 + x = 180 \rightarrow 96 + x = 180 \rightarrow x = 180 - 96 = 84$

9) Area $= \frac{1}{2}$(base×height), base= 8 and height = 5; Area $= \frac{1}{2}(8 \times 5) = \frac{40}{2} = 20 \, ft^2$

10) Base = 7 m and height = 12 m; Area $= \frac{1}{2}(12 \times 7) = \frac{84}{2} = 42 \, m^2$

11) Base = 12 cm and height = 6 cm; Area $= \frac{1}{2}(12 \times 6) = \frac{72}{2} = 36 \, cm^2$

12) Base = 16 in and height = 12 in; Area $= \frac{1}{2}(16 \times 12) = \frac{192}{2} = 96 \, in^2$

13) The perimeter of a parallelogram $= 2(l + w) \rightarrow l = 14 \, cm, w = 16 \, cm$.
    Then: $2(l + w) = 2(14 + 16) = 60 \, cm$

14) The perimeter of a rectangle $= 2(w + l) \rightarrow l = 10 \, ft, w = 8 \, ft$.
    Then: $2(l + w) = 2(10 + 8) = 36 \, ft$

15) Circumference of a circle $= 2\pi r \rightarrow r = 9 \, in$. Then: $2\pi \times 9 = 18 \times 3.14 = 56.52 \, in$

16) The perimeter of a regular hexagon $= 6a \rightarrow a = 7 \, m$. Then: $6 \times 7 = 42 \, m$

17) The area of a trapezoid $= \frac{1}{2}h(b_1 + b_2) \rightarrow b_1 = 12\,m$, $b_2 = 15\,m$ and $h = 8m$.

Then: $A = \frac{1}{2}(8)(12 + 15) = 108\,m^2$

18) $b_1 = 10\,cm$, $b_2 = 14\,cm$ and $h = 6\,cm$. Then: $A = \frac{1}{2}(6)(10 + 14) = 72\,cm^2$

19) $b_1 = 9\,ft$, $b_2 = 13\,ft$ and $h = 7\,ft$. Then: $A = \frac{1}{2}(7)(9 + 13) = 77\,ft^2$

20) $b_1 = 8\,cm$, $b_2 = 12\,cm$ and $h = 6\,cm$. Then: $A = \frac{1}{2}(6)(8 + 12) = 60\,cm^2$

21) The volume of a cube $= (one\,side)^3 = (5)^3 = 125\,cm^3$

22) The volume of a cube $= (one\,side)^3 = (30)^3 = 27{,}000\,ft^3$

23) The volume of a cube $= (one\,side)^3 = (11)^3 = 1{,}331\,in^3$

24) The volume of a cube $= (one\,side)^3 = (7)^3 = 343\,miles^3$

25) The volume of a Rectangular prism $= l \times w \times h \rightarrow l = 9\,cm, w = 5\,cm, h = 8\,cm$.

Then: $V = 9 \times 5 \times 8 = 360\,cm^3$

26) $V = l \times w \times h \rightarrow l = 14\,m, w = 6\,m, h = 9\,m$. Then: $V = 14 \times 6 \times 9 = 756\,m^3$

27) $V = l \times w \times h \rightarrow l = 12\,in, w = 5\,in, h = 8\,in$. Then: $V = 12 \times 5 \times 8 = 480\,in^3$

28) Volume of a Cylinder $= \pi(r)^2 \times h \rightarrow r = 6\,cm, h = 14\,cm$

Then: $\pi(6)^2 \times 14 = 3.14 \times 36 \times 14 = 1{,}582.56 \approx 1{,}582.6$

29) Volume of a Cylinder $= \pi(r)^2 \times h \rightarrow r = 9\,m, h = 12\,m$

Then: $\pi(9)^2 \times 12 = 3.14 \times 81 \times 12 = 3{,}052.08 \approx 3{,}052.1$

30) Volume of a Cylinder $= \pi(r)^2 \times h \rightarrow r = 6\,cm, h = 10\,cm$

Then: $\pi(6)^2 \times 10 = 3.14 \times 36 \times 10 = 1{,}130$

DAY 10 Statistics and Functions

Math topics that you'll learn in this chapter:

1. Mean, Median, Mode, and Range of the Given Data
2. Pie Graph
3. Probability Problems
4. Permutations and Combinations
5. Function Notation and Evaluation
6. Adding and Subtracting Functions
7. Multiplying and Dividing Functions
8. Compositions of Functions

AFOQT Math in 10 Days! | DAY 10 | Statistics and Functions
Mean, Median, Mode, and Range of the
Given Data

Mean, Median, Mode, and Range of the Given Data

☆ **Mean:** $\dfrac{sum\ of\ the\ data}{total\ number\ of\ data\ entires}$

☆ **Mode:** the value in the list that appears most often

☆ **Median:** is the middle number of a group of numbers arranged in order by size.

☆ **Range:** the difference of the largest value and smallest value in the list

Examples:

Example 1. What is the mode of these numbers? $4, 7, 8, 7, 8,\ 9,\ 8,\ 5$

Solution: Mode: the value in the list that appears most often.
Therefore, the mode is number 8. There are three number 8 in the data.

Example 2. What is the median of these numbers? $6, 11, 15, 10, 17, 20, 7$

Solution: Write the numbers in order: $6, 7, 10, 11, 15, 17, 20$
The median is the number in the middle. Therefore, the median is 11.

Example 3. What is the mean of these numbers? $8, 5, 3, 7, 6, 4, 9$

Solution: Mean: $\dfrac{sum\ of\ the\ data}{total\ number\ of\ data\ entires} = \dfrac{8+5+3+7+6+4+9}{7} = \dfrac{42}{7} = 6$

Example 4. What is the range in this list? $9, 2, 5, 10, 15, 22, 7$

Solution: Range is the difference of the largest value and smallest value in the list. The largest value is 22 and the smallest value is 2.
Then: $22 - 2 = 20$

Pie Graph

☆ A Pie Graph (Pie Chart) is a circle chart divided into sectors, each sector represents the relative size of each value.

☆ Pie charts represent a snapshot of how a group is broken down into smaller pieces.

Example:

A library has 650 books that include Mathematics, Physics, Chemistry, English and History. Use the following graph to answer the questions.

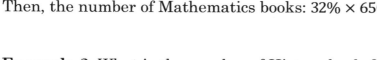

Example 1. What is the number of Mathematics books?

Solution: Number of total books = 650
Percent of Mathematics books = 32%
Then, the number of Mathematics books: 32% × 650 = 0.32 × 650 = 208

Example 2. What is the number of History books?

Solution: Number of total books = 650
Percent of History books = 10%
Then: 0.10 × 650 = 65

Example 3. What is the number of English books in the library?

Solution: Number of total books = 650
Percent of English books = 14%
Then: 0.14 × 650 = 91

Probability Problems

☆ Probability is the likelihood of something happening in the future. It is expressed as a number between zero (can never happen) to 1 (will always happen).

☆ Probability can be expressed as a fraction, a decimal, or a percent.

☆ Probability formula: $Probability = \frac{number\ of\ desired\ outcomes}{number\ of\ total\ outcomes}$

Examples:

Example 1. Anita's trick–or–treat bag contains 8 pieces of chocolate, 16 suckers, 22 pieces of gum and 20 pieces of licorice. If she randomly pulls a piece of candy from her bag, what is the probability of her pulling out a piece of gum?

Solution: $Probability = \frac{number\ of\ desired\ outcomes}{number\ of\ total\ outcomes}$

Probability of pulling out a piece of gum $= \frac{22}{8+16+22+20} = \frac{22}{66} = \frac{1}{3}$

Example 2. A bag contains 25 balls: five green, eight black, seven blue, a brown, a red and three white. If 24 balls are removed from the bag at random, what is the probability that a red ball has been removed?

Solution: If 24 balls are removed from the bag at random, there will be one ball in the bag. The probability of choosing a red ball is 1 out of 25. Therefore, the probability of not choosing a red ball is 24 out of 25 and the probability of having not a red ball after removing 24 balls is the same. The answer is: $\frac{24}{25}$

Permutations and Combinations

☆ **Factorials** are products, indicated by an exclamation mark. For example, $4! = 4 \times 3 \times 2 \times 1$ (Remember that 0! is defined to be equal to 1)

☆ **Permutations:** The number of ways to choose a sample of k elements from a set of n distinct objects where order does matter, and replacements are not allowed. For a permutation problem, use this formula:

$$nP_k = \frac{n!}{(n-k)!}$$

☆ **Combination:** The number of ways to choose a sample of r elements from a set of n distinct objects where order does not matter, and replacements are not allowed. For a combination problem, use this formula:

$$nC_r = \frac{n!}{r!\,(n-r)!}$$

Examples:

Example 1. How many ways can the first and second place be awarded to 6 people?

Solution: Since the order matters, (the first and second place are different!) we need to use permutation formula where n is 6 and k is 2.
Then: $\frac{n!}{(n-k)!} = \frac{6!}{(6-2)!} = \frac{6!}{4!} = \frac{6\times5\times4!}{4!}$, remove 4! from both sides of the fraction. Then: $\frac{6\times5\times4!}{4!} = 6 \times 5 = 30$

Example 2. How many ways can we pick a team of 4 people from a group of 9?

Solution: Since the order doesn't matter, we need to use a combination formula where n is 9 and r is 4.
Then: $\frac{n!}{r!\,(n-r)!} = \frac{9!}{4!\,(9-4)!} = \frac{9!}{4!\,(5)!} = \frac{9\times8\times7\times6\times5!}{4!\,(5)!} = \frac{9\times8\times7\times6}{4\times3\times2\times1} = \frac{3,024}{24} = 126$

Function Notation and Evaluation

☆ Functions are mathematical operations that assign unique outputs to given inputs.

☆ Function notation is the way a function is written. It is meant to be a precise way of giving information about the function without a rather lengthy written explanation.

☆ The most popular function notation is $f(x)$ which is read "f of x". Any letter can name a function. for example: $g(x)$, $h(x)$, etc.

☆ To evaluate a function, plug in the input (the given value or expression) for the function's variable (place holder, x).

Examples:

Example 1. Evaluate: $f(x) = 2x + 9$, find $f(5)$

Solution: Substitute x with 5:
Then: $f(x) = 2x + 9 \rightarrow f(5) = 2(5) + 9 \rightarrow f(5) = 19$

Example 2. Evaluate: $w(x) = 5x - 4$, find $w(2)$.

Solution: Substitute x with 2:
Then: $w(x) = 5x - 4 \rightarrow w(2) = 5(2) - 4 = 10 - 4 = 6$

Example 3. Evaluate: $f(x) = 5x^2 + 8$, find $f(-2)$.

Solution: Substitute x with -2:
Then: $f(x) = 5x^2 + 8 \rightarrow f(-2) = 5(-2)^2 + 8 \rightarrow f(-2) = 20 + 8 = 28$

Example 4. Evaluate: $h(x) = 3x^2 - 4$, find $h(3a)$.

Solution: Substitute x with $3a$:
Then: $h(x) = 3x^2 - 4 \rightarrow h(3a) = 3(3a)^2 - 4 \rightarrow h(3a) = 3(9a^2) - 4 = 27a^2 - 4$

Adding and Subtracting Functions

☆ Just like we can add and subtract numbers and expressions, we can add or subtract functions and simplify or evaluate them. The result is a new function.

☆ For two functions $f(x)$ and $g(x)$, we can create two new functions:

$$(f + g)(x) = f(x) + g(x) \text{ and } (f - g)(x) = f(x) - g(x)$$

Examples:

Example 1. $g(x) = 4x - 3$, $f(x) = x + 6$, Find: $(g + f)(x)$

Solution: $(g + f)(x) = g(x) + f(x)$
Then: $(g + f)(x) = (4x - 3) + (x + 6) = 4x - 3 + x + 6 = 5x + 3$

Example 2. $f(x) = 2x - 7$, $g(x) = x - 9$, Find: $(f - g)(x)$

Solution: $(f - g)(x) = f(x) - g(x)$
Then: $(f - g)(x) = (2x - 7) - (x - 9) = 2x - 7 - x + 9 = x + 2$

Example 3. $g(x) = 2x^2 + 6$, $f(x) = x - 3$, Find: $(g + f)(x)$

Solution: $(g + f)(x) = g(x) + f(x)$
Then: $(g + f)(x) = (2x^2 + 6) + (x - 3) = 2x^2 + x + 3$

Example 4. $f(x) = 2x^2 - 1$, $g(x) = 4x + 3$, Find: $(f - g)(2)$

Solution: $(f - g)(x) = f(x) - g(x)$
Then: $(f - g)(x) = (2x^2 - 1) - (4x + 3) = 2x^2 - 1 - 4x - 3 = 2x^2 - 4x - 4$
Substitute x with 2: $(f - g)(2) = 2(2)^2 - 4(2) - 4 = 8 - 8 - 4 = -4$

Multiplying and Dividing Functions

☆ Just like we can multiply and divide numbers and expressions, we can multiply and divide two functions and simplify or evaluate them.

☆ For two functions $f(x)$ and $g(x)$, we can create two new functions:

$$(f.g)(x) = f(x).g(x) \text{ and } \left(\frac{f}{g}\right)(x) = \frac{f(x)}{g(x)}$$

Examples:

Example 1. $g(x) = x + 2$, $f(x) = x + 5$, Find: $(g.f)(x)$

Solution:

$(g.f)(x) = g(x).f(x) = (x + 2)(x + 5) = x^2 + 5x + 2x + 10 = x^2 + 7x + 10$

Example 2. $f(x) = x + 4$, $h(x) = x - 16$, Find: $\left(\frac{f}{h}\right)(x)$

Solution: $\left(\frac{f}{h}\right)(x) = \frac{f(x)}{h(x)} = \frac{x+4}{x-16}$

Example 3. $g(x) = x + 5$, $f(x) = x - 2$, Find: $(g.f)(3)$

Solution: $(g.f)(x) = g(x).f(x) = (x + 5)(x - 2) = x^2 - 2x + 5x - 10$

$$g(x).f(x) = x^2 + 3x - 10$$

Substitute x with 3: $(g.f)(3) = (3)^2 + 3(3) - 10 = 9 + 9 - 10 = 8$

Example 4. $f(x) = 2x + 2$, $h(x) = x - 3$, Find: $\left(\frac{f}{h}\right)(4)$

Solution: $\left(\frac{f}{h}\right)(x) = \frac{f(x)}{h(x)} = \frac{2x+2}{x-3}$

Substitute x with 4: $\left(\frac{f}{h}\right)(4) = \frac{2x+2}{x-3} = \frac{2(4)+2}{4-3} = \frac{10}{1} = 10$

Composition of Functions

☆ "Composition of functions" simply means combining two or more functions in a way where the output from one function becomes the input for the next function.

☆ The notation used for composition is: $(fog)(x) = f(g(x))$ and is read "f composed with g of x" or "f of g of x".

Examples:

Example 1. Using $f(x) = x + 5$ and $g(x) = 7x$, find: $(fog)(x)$

Solution: $(fog)(x) = f(g(x))$. Then: $(fog)(x) = f(g(x)) = f(7x)$

Now find $f(7x)$ by substituting x with $7x$ in $f(x)$ function.

Then: $f(x) = x + 5$; $(x \rightarrow 7x) \rightarrow f(7x) = 7x + 5$

Example 2. Using $f(x) = 2x - 3$ and $g(x) = x - 5$, find: $(gof)(3)$

Solution: $(fog)(x) = f(g(x))$. Then: $(gof)(x) = g(f(x)) = g(2x - 3)$,

Now substitute x in $g(x)$ by $(2x - 3)$.

Then: $g(2x - 3) = (2x - 3) - 5 = 2x - 8$

Substitute x with 3: $(gof)(3) = g(f(x)) = 2x - 8 = 2(3) - 8 = -2$

Example 3. Using $f(x) = 3x^2 - 7$ and $g(x) = 2x + 1$, find: $f(g(2))$

Solution: First, find $g(2)$: $g(x) = 2x + 1 \rightarrow g(2) = 2(2) + 1 = 5$

Then: $f(g(2)) = f(5)$. Now, find $f(5)$ by substituting x with 5 in $f(x)$ function.

$f(g(2)) = f(5) = 3(5)^2 - 7 = 3(25) - 7 = 68$

Day 10: Practices

✍ Find the values of the Given Data.

1) $9, 10, 9, 8, 11$

2) $6, 8, 1, 4, 7, 6, 10$

Mode: _____   Range: _____   Mean: _____   Median: _____

✍ The circle graph below shows all Bob's expenses for last month. Bob spent $675 on his Rent last month.

Bob's last month expenses

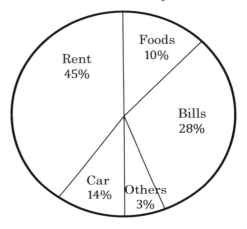

3) How much did Bob's total expenses last month? _____

4) How much did Bob spend for his bills last month? _____

5) How much did Bob spend for his car last month? _____

✍ Solve.

6) Bag A contains 9 red marbles and 6 green marbles. Bag B contains 6 black marbles and 9 orange marbles. What is the probability of selecting a red marble at random from bag A? What is the probability of selecting a black marble at random from Bag B?

_____   _____

7) Jason is planning for his vacation. He wants to go to museum, go to the beach, and play volleyball. How many different ways of ordering are there for him? _____

8) In how many ways can a team of 8 basketball players choose a captain and co-captain? _____

9) How many ways can you give 9 balls to your 12 friends? _____

✏ Evaluate each function.

10) $g(n) = 3n + 7$, find $g(3)$

_____

11) $h(x) = 4n - 7$, find $h(5)$

_____

12) $y(n) = 12 - 3n$, find $y(8)$

_____

13) $b(n) = -10 - 5n$, find $b(8)$

_____

14) $g(x) = -7x + 6$, find $g(-3)$

_____

15) $k(n) = -4n + 5$, find $k(-5)$

_____

16) $w(n) = -3n - 6$, find $w(-4)$

_____

17) $z(n) = 14 - 2n$, find $n(3)$

_____

✏ Perform the indicated operation.

18) $f(x) = 2x + 3$

$g(x) = x + 4$

Find $(f - g)(x)$

_____

19) $g(a) = -3a + 2$

$h(a) = a^2 - 4$

Find $(h + g)(3)$

_____

✏ Perform the indicated operation.

20) $g(x) = x - 4$

$f(x) = x + 3$

Find $(g.f)(2)$

_____

21) $f(x) = x + 2$

$h(x) = x - 5$

Find$\left(\frac{f}{h}\right)(-3)$

_____

✏ Using $f(x) = 2x + 5$ and $g(x) = 2x - 3$, find:

22) $g\big(f(2)\big) = $ _____

23) $g\big(f(-2)\big) = $ _____

24) $f\big(f(1)\big) = $ _____

25) $f\big(f(-1)\big) = $ _____

26) $f\big(g(4)\big) = $ _____

27) $f\big(g(-3)\big) = $ _____

Day 10: Answers

1) Mode = the value in the list that appears most often = 9

Range: the difference of the largest value and smallest value. Then:
$$11 - 8 = 3$$

2) Mean: $\dfrac{sum\ of\ the\ data}{total\ number\ of\ data\ entires} = \dfrac{6+8+1+4+7+6+10}{7} = \dfrac{42}{7} = 6$

Median: is the middle number of a group of numbers arranged in order by size. Write the numbers in order: 1, 4, 6, 7, 8, 10. Then: Median $= \dfrac{6+7}{2} = 6.5$

3) Bob's rent = $675

Total expenses → 45% of total expenses = rent →

Total expenses = rent ÷ 0.45 = $675 ÷ 0.45 = $1,500

4) Bills = 28% × total expenses = 0.28 × $1,500 = $420

5) Car = 14% × total expenses = 0.14 × $1,500 = 210

6) Probability of pulling out a red marble from bag A $= \dfrac{numbers\ of\ red\ marbles}{total\ of\ marbles} =$

$\dfrac{9}{6+9} = \dfrac{9}{15} = 0.6 = 60\%$

Probability of pulling out a black marble from bag B =
$\dfrac{numbers\ of\ black\ marbles}{total\ of\ marbles} = \dfrac{6}{6+9} = \dfrac{6}{15} = 0.4 = 40\%$

7) Jason has 3 choices. Therefore, number of different ways of ordering the events is: $3 \times 2 \times 1 = 6$

8) This is a permutation problem. (order is important) Then: $nP_k = \dfrac{n!}{(n-k)!} \to n = 8, k = 2 \to \dfrac{8!}{(8-2)!} = \dfrac{8!}{6!} = \dfrac{8 \times 7 \times 6!}{6!} = 56$

9) This is a combination problem (order doesn't matter) $nC_r = \dfrac{n!}{r!\,(n-r)!} \to$
$$n = 12, r = 9 \to nC_r = \dfrac{12!}{9!\,(12-9)!} = \dfrac{12!}{9!\,3!} = \dfrac{12 \times 11 \times 10 \times 9!}{3 \times 2 \times 1 \times 9!} = \dfrac{1,320}{6} = 220$$

10) $g(n) = 3n + 7 \to g(3) = 3(3) + 7 = 9 + 7 = 16$

11) $h(x) = 4n - 7 \to h(5) = 4(5) - 7 = 20 - 7 = 13$

12) $y(n) = 12 - 3n \to y(8) = 12 - 3(8) = 12 - 24 = -12$

13) $b(n) = -10 - 5n \to b(8) = -10 - 5(8) = -10 - 40 = -50$

14) $g(x) = -7x + 6 \to g(-3) = -7(-3) + 6 = 21 + 6 = 27$

15) $k(n) = -4n + 5 \to k(-5) = -4(-5) + 5 = 20 + 5 = 25$

16) $w(n) = -3n - 6 \rightarrow w(-4) = -3(-4) - 6 = 12 - 6 = 6$

17) $z(n) = 14 - 2n \rightarrow n(3) = 14 - 2(3) = 14 - 6 = 8$

18) $(f - g)(x) = f(x) - g(x) = 2x + 3 - (x + 4) = 2x + 3 - x - 4 = x - 1$

19) $(h + g)(3) = h(a) + g(a) = a^2 - 4 - 3a + 2 = a^2 - 3a - 2 = (3)^2 - 3(3) - 2 =$
$9 - 9 - 2 = -2$

20) $(g.f)(2) = f(x) \times g(x) = (x - 4) \times (x + 3) = (x \times x) + (x \times 3) + (-4 \times x) +$
$(-4 \times 3) = x^2 + 3x - 4x - 12 = x^2 - x - 12 = 2^2 - 2 - 12 = 4 - 2 - 12 = -10$

21) $\left(\frac{f}{h}\right)(-3) = \frac{f(x)}{h(x)} = \frac{x+2}{x-5} = \frac{(-3)+2}{(-3)-5} = \frac{-3+2}{-3-5} = \frac{-1}{-8} = \frac{1}{8}$

22) First, find $f(2)$: $f(x) = 2x + 5 \rightarrow f(2) = 2(2) + 5 = 9$

Then: $g(f(2)) = g(9)$. Now, find $g(9)$ by substituting x with 9 in $g(x)$

function. $g(f(2)) = g(9) = 2(9) - 3 = 18 - 3 = 15$

23) First, find $f(-2)$: $f(x) = 2x + 5 \rightarrow f(-2) = 2(-2) + 5 = -4 + 5 = 1$

Then: $g(f(-2)) = g(1)$. Now, find $g(1)$ by substituting x with 1 in $g(x)$

function. $g(f(-2)) = g(1) = 2(1) - 3 = 2 - 3 = -1$

24) First, find $f(1)$: $f(x) = 2x + 5 \rightarrow f(1) = 2(1) + 5 = 2 + 5 = 7$

Then: $f(f(1)) = f(7)$. Now, find $f(7)$ by substituting x with 7 in $f(x)$

function. $f(f(1)) = f(7) = 2(7) + 5 = 14 + 5 = 19$

25) First, find $f(-1)$: $f(x) = 2x + 5 \rightarrow f(-1) = 2(-1) + 5 = -2 + 5 = 3$

Then: $f(f(-1)) = f(3)$. Now, find $f(3)$ by substituting x with 3 in $f(x)$

function. $f(g(-1)) = f(3) = 2(3) + 5 = 6 + 5 = 11$

26) First, find $g(4)$: $g(x) = 2x - 3 \rightarrow g(4) = 2(4) - 3 = 8 - 3 = 5$

Then: $f(g(4)) = f(5)$. Now, find $f(5)$ by substituting x with 5 in $f(x)$

function. $f(g(4)) = f(5) = 2(5) + 5 = 10 + 5 = 15$

27) First, find $g(-3)$: $g(x) = 2x - 3 \rightarrow g(-3) = 2(-3) - 3 = -6 - 3 = -9$

Then: $f(g(-3)) = f(-9)$. Now, find $f(-9)$ by substituting x with -9 in

$f(x)$. function. $f(g(-3)) = f(-9) = 2(-9) + 5 = -18 + 5 = -13$

AFOQT Test Review

The Air Force Officer Qualifying Test (AFOQT) is a standardized test to assess skills and personality traits that have proven to be predictive of success in officer commissioning programs such as the training program.

The AFOQT is used to select applicants for officer commissioning programs, such as Officer Training School (OTS) or Air Force Reserve Officer Training Corps (Air Force ROTC) and pilot and navigator training.

The AFOQT is a multiple-aptitude battery that measures developed abilities and helps predict future academic and occupational success in the military. The AFOQT is a multiple-choice test which consists of 12 subtests and two of them are Arithmetic Reasoning and Mathematics Knowledge.

In this section, there are 2 complete Arithmetic Reasoning and Mathematics Knowledge AFOQT Tests. Take these tests to see what score you'll be able to receive on a real AFOQT test.

Good luck!

Time to refine your skill with a practice examination

Take AFOQT Arithmetic Reasoning and Mathematics Knowledge tests to simulate the test day experience. After you've finished, score your tests using the answer keys.

Before You Start

- ❖ You'll need a pencil and a timer to take the test.
- ❖ For each question, there are five possible answers. Choose which one is best.
- ❖ It's okay to guess. There is no penalty for wrong answers.
- ❖ Use the answer sheet provided to record your answers.
- ❖ After you've finished the test, review the answer key to see where you went wrong.

Calculators are NOT permitted for the AFOQT Test

Good Luck!

AFOQT Math

Practice Test 1

2020 – 2021

Section 1: Arithmetic Reasoning

Total time for this section: 29 Minutes

25 questions

You may NOT use a calculator on this Section.

1) What is 5 percent of 480?
   A. 20
   B. 24
   C. 30
   D. 40
   E. 50

2) In two successive years, the population of a town is increased by 15% and 20%. What percent of the population is increased after two years?
   A. 32%
   B. 35%
   C. 38%
   D. 68%
   E. 86%

3) The marked price of a computer is D dollar. Its price decreased by 25% in January and later increased by 10% in February. What is the final price of the computer in D dollar?
   A. 0.80 D
   B. 0.82 D
   C. 0.90 D
   D. 1.20 D
   E. 1.50 D

4) Last week 24,000 fans attended a football match. This week three times as many bought tickets, but one sixth of them cancelled their tickets. How many are attending this week?
   A. 48,000
   B. 54,000
   C. 60,000
   D. 72,000
   E. 84,000

5) The average of 13, 15, 20 and x is 20. What is the value of x?
   A. 9
   B. 15
   C. 18
   D. 32
   E. 64

6) In 1999, the average worker's income increased $2,000 per year starting from $26,000 annual salary. Which equation represents income greater than average? (I = income, x = number of years after 1999)

A. $I > 2000\,x + 26000$

B. $I > -2000\,x + 26000$

C. $I < -2000\,x + 26000$

D. $I < 2000\,x - 26000$

E. $I < -2000\,x - 26000$

7) Jason deposits 15% of $160 into a savings account, what is the amount of his deposit?

A. $10

B. $16

C. $20

D. $24

E. $30

8) If 150% of a number is 75, then what is the 90% of that number?

A. 45

B. 50

C. 70

D. 85

E. 90

9) What is the remainder when 1,454 is divided by 7?

A. 2

B. 3

C. 5

D. 6

E. 8

10) The score of Emma was half as that of Ava and the score of Mia was twice that of Ava. If the score of Mia was 40, what is the score of Emma?

A. 10

B. 15

C. 20

D. 30

E. 40

11) Mr. Jones saves $2,500 out of his monthly family income of $55,000. What fractional part of his income does he save?

A. $\frac{1}{22}$

B. $\frac{1}{11}$

C. $\frac{3}{25}$

D. $\frac{2}{15}$

E. $\frac{2}{25}$

12) 16% of what number is equal to 72?

A. 8.64

B. 36

C. 300

D. 350

E. 450

13) 55 students took an exam and 11 of them failed. What percent of the students passed the exam?

A. 20%

B. 40%

C. 60%

D. 75%

E. 80%

14) A bag contains 18 balls: two green, five black, eight blue, a brown, a red and one white. If 17 balls are removed from the bag at random, what is the probability that a brown ball has been removed?

A. $\frac{1}{9}$

B. $\frac{1}{6}$

C. $\frac{16}{11}$

D. $\frac{17}{18}$

E. $\frac{22}{25}$

15) What is 0.5749 rounded to the nearest hundredth?
   A. 0.57
   B. 0.575
   C. 0.58
   D. 0.584
   E. 0.585

16) If a gas tank can hold 28 gallons, how many gallons does it contain when it is $\frac{3}{4}$ full?
   A. 84
   B. 168
   C. 33.5
   D. 21
   E. 12

17) Ethan needs an 75% average in his writing class to pass. On his first 4 exams, he earned scores of 68%, 72%, 85%, and 90%. What is the minimum score Ethan can earn on his fifth and final test to pass?
   A. 80%
   B. 70%
   C. 68%
   D. 60%
   E. 50%

18) A chemical solution contains 6% alcohol. If there is 24 ml of alcohol, what is the volume of the solution?
   A. 240 ml
   B. 400 ml
   C. 600 ml
   D. 1,200 ml
   E. 1,800 ml

19) The average of five consecutive numbers is 36. What is the smallest number?
   A. 38
   B. 36
   C. 34
   D. 12
   E. 8

20) Which of the following is equivalent to $\frac{2}{5}$?
   A. 0.04
   B. 0.25
   C. 0.40
   D. 1.4
   E. 1.65

21) A tree 32 feet tall casts a shadow 12 feet long. Jack is 6 feet tall. How long is Jack's shadow?
   A. 2.25 *feet*
   B. 4 *feet*
   C. 4.25 *feet*
   D. 8 *feet*
   E. 8.25 *feet*

22) 15% of what number is equal to 75?
   A. 8.64
   B. 36
   C. 300
   D. 500
   E. 700

23) Five years ago, Amy was three times as old as Mike was. If Mike is 10 years old now, how old is Amy?

   A. 4
   B. 8
   C. 12
   D. 20
   E. 24

24) John traveled 150 km in 6 hours and Alice traveled 140 km in 4 hours. What is the ratio of the average speed of John to average speed of Alice?

    A. 3 : 2

    B. 2 : 3

    C. 5 : 7

    D. 5 : 6

    E. 7 : 5

25) 12 less than twice a positive integer is 80. What is the integer?

    A. 46

    B. 48

    C. 50

    D. 55

    E. 60

STOP: This is the End of Section 1 of test 1.

AFOQT Math

Practice Test 1

2020 – 2021

Section 2:
Mathematics Knowledge

Total time for this section: 22 Minutes

25 questions

You may NOT use a calculator on this Section.

1) Which of the following is equal to the expression below?

$$(5x + 2y)(2x - y)$$

A. $4x^2 - 2y^2$
B. $2x^2 + 6xy - 2y^2$
C. $24x^2 + 2xy - 2y^2$
D. $10x^2 - xy - 2y^2$
E. $15x^2 + xy - 2y^2$

2) What is the product of all possible values of x in the following equation?

$$|x - 10| = 4$$

A. 3
B. 7
C. 13
D. 64
E. 84

3) What is the slope of a line that is perpendicular to the line $4x - 2y = 6$?
A. -2
B. $-\frac{1}{2}$
C. 4
D. 12
E. 16

4) What is the value of the expression $6(x - 2y) + (2 - x)^2$ when $x = 3$ and $y = -2$?
A. -4
B. 20
C. 43
D. 50
E. 64

5) What is the area of á square whose diagonal is 4?
   A. 4
   B. 8
   C. 16
   D. 64
   E. 98

6) What is the value of x in the following equation? $\frac{2}{3}x + \frac{1}{6} = \frac{1}{2}$
   A. 6
   B. $\frac{1}{2}$
   C. $\frac{1}{3}$
   D. $\frac{1}{4}$
   E. $\frac{3}{4}$

7) A bank is offering 4.5% simple interest on a savings account. If you deposit $12,000, how much interest will you earn in two years?
   A. $420
   B. $1,080
   C. $4,200
   D. $8,400

8) Simplify $7x^2y^3(2x^2y)^3 =$

   A. $12x^4y^6$
   B. $12x^8y^6$
   C. $56x^4y^6$
   D. $56x^8y^6$
   E. $65x^6y^8$

9) What is the surface area of the cylinder below?
   A. $40\,\pi\ in^2$
   B. $57\,\pi\ in^2$
   C. $66\,\pi\ in^2$
   D. $288\,\pi\ in^2$
   E. $322\,\pi\ in^2$

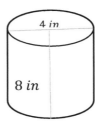

4 in

8 in

10) A cruise line ship left Port A and traveled 50 miles due west and then 120 miles due north. At this point, what is the shortest distance from the cruise to port A?

A. 70 *miles*
B. 80 *miles*
C. 150 *miles*
D. 130 *miles*
E. 160 *miles*

11) What is the equivalent temperature of 104°F in Celsius?

$$C = \frac{5}{9}(F - 32)$$

A. 32
B. 40
C. 48
D. 52
E. 56

12) The perimeter of a rectangular yard is 72 meters. What is its length if its width is twice its length?

A. 12 *meters*
B. 18 *meters*
C. 20 *meters*
D. 24 *meters*
E. 32 *meters*

13) What is the slope of the line: $4x - 2y = 12$

A. -1
B. -2
C. 1
D. 2
E. -4

14) The area of a circle is 36π. What is the diameter of the circle?
  A. 4
  B. 8
  C. 12
  D. 14
  E. 16

15) If $f(x) = 2x^3 + 5x^2 + 2x$ and $g(x) = -3$, what is the value of $f(g(x))$?
  A. 36
  B. 32
  C. 24
  D. -15
  E. -20

16) The diagonal of a rectangle is 10 inches long and the height of the rectangle is 6 inches. What is the perimeter of the rectangle?
  A. 10 *inches*
  B. 12 *inches*
  C. 16 *inches*
  D. 28 *inches*
  E. 40 *inches*

17) The perimeter of the trapezoid below is 40 *cm*. What is its area?
  A. $48 \ cm^2$
  B. $98 \ cm^2$
  C. $140 \ cm^2$
  D. $576 \ cm^2$
  E. $686 \ cm^2$

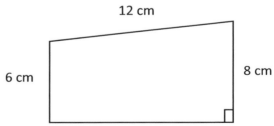

18) If $f(x)=2x^3+ 2$ and $(x) = \frac{1}{x}$, what is the value of $f(g(x))$?

  A. $\frac{1}{2x^3+2}$
  B. $\frac{2}{x^3}$
  C. $\frac{1}{2x}$
  D. $\frac{2}{x^3} + 2$
  E. $\frac{2}{x^3+2}$

19) A cruise line ship left Port A and traveled 80 miles due west and then 150 miles due north. At this point, what is the shortest distance from the cruise to port A?

A. 70 *miles*
B. 80 *miles*
C. 150 *miles*
D. 170 *miles*
E. 240 *miles*

20) If the ratio of $5a$ to $2b$ is $\frac{1}{10}$, what is the ratio of a to b?

A. 10
B. 25
C. $\frac{1}{25}$
D. $\frac{1}{20}$
E. $\frac{1}{15}$

21) If $x = 9$, what is the value of y in the following equation? $2y = \frac{2x^2}{3} + 6$

A. 30
B. 45
C. 60
D. 120
E. 140

22) If $\frac{x-3}{5} = N$ and $N = 6$, what is the value of x?
A. 25
B. 28
C. 30
D. 33
E. 35

23) Which of the following is equal to $b^{\frac{3}{5}}$?

A. $\sqrt{b^{\frac{5}{3}}}$

B. $b^{\frac{5}{3}}$

C. $\sqrt[5]{b^3}$

D. $\sqrt[3]{b^5}$

E. $\sqrt[5]{b^8}$

24) On Saturday, Sara read N pages of a book each hour for 3 hours, and Mary read M pages of a book each hour for 4 hours. Which of the following represents the total number of pages of book read by Sara and Mary on Saturday?

A. $12MN$

B. $3N + 4M$

C. $7MN$

D. $4N + 3M$

E. $5N + 4M$

25) Sara opened a bank account that earns 2 percent compounded annually. Her initial deposit was \$150, and she uses the expression $\$150(x)^n$ to find the value of the account after n years. What is the value of x in the expression?

A. 0.02

B. 0.20

C. 20%

D. 1

E. 1.02

STOP: This is the End of Section 2 of test 1.

AFOQT Math

Practice Test 2

2020 – 2021

Section 1: Arithmetic Reasoning

Total time for this section: 29 Minutes

25 questions

You may NOT use a calculator on this Section.

1) Will has been working on a report for 6 hours each day, 7 days a week for 2 weeks. How many minutes has Will worked on his report?
   A. 42
   B. 84
   C. 2,520
   D. 4,432
   E. 5,040

2) James is driving to visit his mother, who lives 340 miles away. How long will the drive be, round–trip, if James drives at an average speed of 50 mph?
   A. 135 *minutes*
   B. 310 *minutes*
   C. 741 *minutes*
   D. 756 *minutes*
   E. 816 *minutes*

3) In a classroom of 60 students, 42 are female. What percentage of the class is male?
   A. 34%
   B. 22%
   C. 30%
   D. 26%
   E. 38%

4) You are asked to chart the temperature during a 6-hour period to give the average. These are your results:
   7 am: 7 degrees
   8 am: 9 degrees
   9 am: 22 degrees
   10 am: 28 degrees
   11 am: 28 degrees
   12 pm: 30 degrees
   What is the average temperature?
   A. 32.67
   B. 24.67
   C. 20.67
   D. 18.27
   E. 16.27

5) During the last week of track training, Emma achieves the following times in seconds: 66, 57, 54, 64, 57, and 59. Her three best times this week (least times) are averaged for her final score on the course. What is her final score?

A. 56 *seconds*

B. 57 *seconds*

C. 59 *seconds*

D. 61 *seconds*

E. 63 *seconds*

6) How many square feet of tile is needed for a 15 feet x 15 feet room?

A. 225 *square feet*

B. 118.5 *square feet*

C. 112 *square feet*

D. 60 *square feet*

E. 40 *square feet*

7) With what number must 1.303572 be multiplied in order to obtain the number 1303.572?

A. 100

B. 1,000

C. 10,000

D. 100,000

E. 1,000,000

8) Which of the following is NOT a factor of 50?

A. 5

B. 2

C. 10

D. 15

E. 20

9) Emma is working in a hospital supply room and makes $25.00 an hour. The union negotiates a new contract giving each employee a 4% cost of living raise. What is Emma's new hourly rate?

A. $26 *an hour*

B. $28 *an hour*

C. $30 *an hour*

D. $31.50 *an hour*

E. $32.50 *an hour*

10) Emily and Lucas have taken the same number of photos on their school trip. Emily has taken 4 times as many photos as Mia. Lucas has taken 21 more photos than Mia. How many photos has Mia taken?

A. 7

B. 9

C. 11

D. 13

E. 15

11) Will has been working on a report for 5 hours each day, 6 days a week for 2 weeks. How many minutes has Will worked on his report?

A. 7,444 *minutes*

B. 5,524 *minutes*

C. 3,600 *minutes*

D. 2,640 *minutes*

E. 1,300 *minutes*

12) Find the average of the following numbers: $22, 34, 16, 20$

A. 23

B. 35

C. 30

D. 23.3

E. 35.3

13) A mobile classroom is a rectangular block that is 90 feet by 30 feet in length and width respectively. If a student walks around the block once, how many yards does the student cover?

    A. 2,700 *yards*

    B. 240 *yards*

    C. 120 *yards*

    D. 60 *yards*

    E. 30 *yards*

14) What is the distance in miles of a trip that takes 2.1 hours at an average speed of 16.2 miles per hour? (Round your answer to a whole number)

    A. 44 *miles*

    B. 34 *miles*

    C. 30 *miles*

    D. 18 *miles*

    E. 12 *miles*

15) The sum of 6 numbers is greater than 120 and less than 180. Which of the following could be the average (arithmetic mean) of the numbers?

    A. 20

    B. 26

    C. 30

    D. 34

    E. 38

16) A barista averages making 15 coffees per hour. At this rate, how many hours will it take until she's made 1,500 coffees?

    A. 95 *hours*

    B. 90 *hours*

    C. 100 *hours*

    D. 105 *hours*

    E. 115 *hours*

17) There are 120 rooms that need to be painted and only 12 painters available. If there are still 12 rooms unpainted by the end of the day, what is the average number of rooms that each painter has painted?

A. 9

B. 12

C. 14

D. 16

E. 18

18) Nicole was making $7.50 per hour and got a raise to $7.75 per hour. What percentage increase was Nicole's raise?

A. 2%

B. 1.67%

C. 3.33%

D. 6.66%

E. 9.99%

19) An architect's floor plan uses ½ inch to represent one mile. What is the actual distance represented by 4 ½ inches?

A. 9 *miles*

B. 8 *miles*

C. 7 *miles*

D. 6 *miles*

E. 5 *miles*

20) A snack machine accepts only quarters. Candy bars cost 25¢, a package of peanuts costs 75¢, and a can of cola costs 50¢. How many quarters are needed to buy two Candy bars, one package of peanuts, and one can of cola?

A. 8 *quarters*

B. 7 *quarters*

C. 6 *quarters*

D. 5 *quarters*

E. 4 *quarters*

21) The hour hand of a watch rotates 30 degrees every hour. How many complete rotations does the hour hand make in 8 days?

   A. 12

   B. 14

   C. 16

   D. 18

   E. 20

22) What is the product of the square root of 81 and the square root of 25?

   A. 2,025

   B. 15

   C. 25

   D. 35

   E. 45

23) If $2y + 4y + 2y = -24$, then what is the value of y?

   A. -3

   B. -2

   C. -1

   D. 0

   E. 1

24) A bread recipe calls for $2\frac{2}{3}$ cups of flour. If you only have $1\frac{5}{6}$ cups of flour, how much more flour is needed?

   A. 1

   B. $\frac{1}{2}$

   C. 2

   D. $\frac{5}{6}$

   E. 5

25) Convert 0.023 to a percent.
   A. 0.2%
   B. 0.23%
   C. 2.30%
   D. 23%
   E. 23.20%

STOP: This is the End of Section 1 of test 2.

AFOQT Math

Practice Test 2

2020 – 2021

Section 2:
Mathematics Knowledge

Total time for this section: 22 Minutes

25 questions

You may NOT use a calculator on this Section.

1) $(x + 7)(x + 5) = ?$
A. $x^2 + 2x + 12$
B. $x^2 + 12x + 12$
C. $x^2 + 35x + 12$
D. $x^2 + 12x + 35$
E. $x^2 - 12x + 45$

2) Convert 670,000 to scientific notation.
A. 6.70×1000
B. 6.70×10^{-5}
C. 6.70×100
D. 6.7×10^5
E. 67×10^5

3) What is the perimeter of the triangle in the provided diagram?
A. 15,625
B. 625
C. 75
D. 25
E. 15

4) If x is a positive integer divisible by 6, and $x < 60$, what is the greatest possible value of x?
A. 54
B. 48
C. 36
D. 59
E. 65

5) There are two pizza ovens in a restaurant. Oven 1 burns four times as many pizzas as oven 2. If the restaurant had a total of 15 burnt pizzas on Saturday, how many pizzas did oven 2 burn?
A. 3
B. 6
C. 9
D. 12
E. 15

6) Which of the following is an obtuse angle?

A. 56°

B. 72°

C. 123°

D. 211°

E. 243°

7) If $8 + 2x$ is 16 more than 20, what is the value of $6x$?

A. 40

B. 55

C. 62

D. 72

E. 84

8) If a gas tank can hold 30 gallons, how many gallons does it contain when it is $\frac{3}{5}$ full?

A. 27

B. 24

C. 21

D. 20

E. 18

9) In the xy-plane, the point $(4,3)$ and $(3,2)$ are on line A. Which of the following equations of lines is parallel to line A?

A. $y = 3x$

B. $y = \frac{x}{2}$

C. $y = 2x$

D. $y = x$

E. $y = 4x$

10) A circle has a diameter of 16 inches. What is its approximate area? ($\pi = 3.14$)

A. 200.96

B. 100.48

C. 64.00

D. 12.56

E. 10.00

11) If $y = (-3x^3)^2$, which of the following expressions is equal to y?

A. $-6x^5$

B. $-6x^6$

C. $6x^5$

D. $9x^6$

E. $12x^5$

12) The equation of a line is given as: $y = 5x - 3$. Which of the following points does not lie on the line?

A. $(1, 2)$

B. $(-2, -13)$

C. $(3, 18)$

D. $(2, 7)$

E. $(5, 3)$

13) When $P + Q = 12$ and $3R + Q = 12$, what is the value of R?

A. 12

B. 2

C. 1

D. 0

E. It cannot be determined from the information given.

14) What is the distance between the points $(1, 3)$ and $(-2, 7)$?

A. 3

B. 4

C. 5

D. 6

E. 7

15) $x^2 - 81 = 0$, x could be:

A. 6

B. 9

C. 12

D. 15

E. 17

16) A rectangular plot of land is measured to be 160 feet by 200 feet. Its total area is:

A. 32,000 square feet
B. 4,404 square feet
C. 3,200 square feet
D. 2,040 square feet
E. 1,400 square feet

17) In the figure below, line A is parallel to line B. What is the value of angle x?

A. 45 degree
B. 55 degree
C. 80 degree
D. 120 degree
E.   140 degree

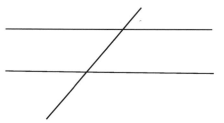

18) What is the length of AB in the following figure if AE = 6, CD = 9 and AC = 24?

A. 6.8
B. 8.2
C. 9.6
D. 18
E. 20

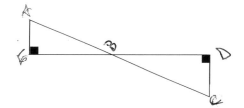

19) John buys a pepper plant that is 6 inches tall. With regular watering the plant grows 4 inches a year. Writing John's plant's height as a function of time, what does the y −intercept represent?

A. The y −intercept represents the rate of grows of the plant which is 4 inches
B. The y −intercept represents the starting height of 6 inches
C. The y −intercept represents the starting height of 4 inches
D. The y −intercept represents the rate of growth of plant which is 3 inches per year
E. There is no y −intercept

20) One fourth the cube of 4 is:

A. 25

B. 16

C. 32

D. 8

E. 2

21) What is the sum of the prime numbers in the following list of numbers?
$$14, 12, 11, 16, 13, 20, 19, 36, 30$$

A. 26

B. 37

C. 43

D. 32

E. 25

22) If $f(x^2) = 3x + 4$, for all positive value of x, what is the value of $f(144)$?

A. 367

B. 41

C. 29

D. −29

E. −41

23) The supplement angle of a 45° angle is:

A. 135°

B. 105°

C. 90°

D. 35°

E. 15°

24) Which of the following is the solution of the following inequality?
$$3.5x - 17.5 < 2x - 5 - 3.5x$$

A. $x < 2.5$
B. $x > -4.11$
C. $x \leq 3$
D. $x \geq -3$
E. $x \leq -4$

25) Simplify: $5(2x^6)^3$.

A. $10x^9$
B. $10x^{18}$
C. $40x^{18}$
D. $40x^9$
E. $60x^{18}$

STOP: This is the End of Section 2 of test 2.

AFOQT Math Practice Tests Answer Keys

Now, it's time to review your results to see where you went wrong and what areas you need to improve.

AFOQT Math Practice Test 1						
Arithmetic Reasoning				**Mathematics Knowledge**		
1)	B	16)	D	1)	D	16)	D
2)	C	17)	D	2)	E	17)	B
3)	B	18)	B	3)	B	18)	D
4)	C	19)	C	4)	C	19)	D
5)	D	20)	C	5)	B	20)	C
6)	A	21)	A	6)	B	21)	A
7)	D	22)	D	7)	B	22)	D
8)	A	23)	D	8)	D	23)	C
9)	C	24)	C	9)	A	24)	B
10)	A	25)	A	10)	D	25)	E
11)	A			11)	B		
12)	E			12)	A		
13)	E			13)	D		
14)	D			14)	C		
15)	A			15)	D		

AFOQT Math Practice Test 2

Arithmetic Reasoning				Mathematics Knowledge			
1)	E	16)	C	1)	D	16)	C
2)	E	17)	A	2)	D	17)	D
3)	C	18)	C	3)	C	18)	C
4)	C	19)	A	4)	A	19)	B
5)	A	20)	B	5)	A	20)	B
6)	A	21)	C	6)	C	21)	C
7)	B	22)	E	7)	E	22)	B
8)	D	23)	A	8)	E	23)	A
9)	A	24)	D	9)	D	24)	A
10)	A	25)	C	10)	A	25)	C
11)	C			11)	D		
12)	A			12)	C		
13)	B			13)	D		
14)	B			14)	C		
15)	B			15)	B		

AFOQT Math Practice Tests Answers and Explanations

AFOQT Practice Test 1: Arithmetic Reasoning

1) Choice B is correct.

5 percent of 480 $= \frac{5}{100} \times 480 = \frac{1}{20} \times 480 = \frac{480}{20} = 24$

2) Choice C is correct

The population is increased by 15% and 20%. 15% increase changes the population to 115% of original population. For the second increase, multiply the result by 120%.

$(1.15) \times (1.20) = 1.38 = 138\%$. 38 percent of the population is increased after two years.

3) Choice B is correct

To find the discount, multiply the number by $(100\% - rate\ of\ discount)$.

Therefore, for the first discount we get: $(D)(100\% - 25\%) = (D)(0.75) = 0.75\ D$

For increase of 10%: $(0.75\ D)(100\% + 10\%) = (0.75\ D)(1.10) = 0.82\ D = 82\%\ of\ D$

4) Choice C is correct

Three times of 24,000 is 72,000. One sixth of them cancelled their tickets.

One sixth of 72,000 equals 12,000 $(\frac{1}{6} \times 72,000 = 12,000)$.

$60,000(72,000 - 12,000 = 60,000)$ fans are attending this week

5) Choice D is correct

average $= \frac{\text{sum of terms}}{\text{number of terms}} \Rightarrow 20 = \frac{13+15+20+x}{4} \Rightarrow 80 = 48 + x \Rightarrow x = 32$

6) Choice A is correct

Let x be the number of years. Therefore, \$2,000 per year equals $2000x$. starting from \$26,000 annual salary means you should add that amount to $2000x$. Income more than that is:

$I > 2000\ x + 26000$

7) Choice D is correct

15% of \$160 is $0.15 \times 160 = 24$

8) Choice A is correct

First, find the number. Let x be the number. 150% of a number is 75, then:

$1.5 \times x = 75 \Rightarrow x = 75 \div 1.5 = 50$

90% of 50 is: $0.9 \times 50 = 45$

9) Choice C is correct

The result when 1,454 is divided by 7 is 207 with a remainder of 5. Multiplying $7 \times 207 = 1,449$ and $1,454 - 1,449 = 5$, which is the remainder.

10) Choice A is correct

If the score of Mia was 40, therefore the score of Ava is 20. Since, the score of Emma was half as that of Ava, therefore, the score of Emma is 10.

11) Choice A is correct

2,500 out of 55,000 equals to $\dfrac{2500}{55000} = \dfrac{25}{550} = \dfrac{1}{22}$

12) Choice E is correct

Dividing 72 by 16%, which is equivalent to 0.16, gives 450.

13) Choice E is correct

The failing rate is 11 out of $55 = \dfrac{11}{55}$. Change the fraction to percent: $\dfrac{11}{55} \times 100\% = 20\%$

20 percent of students failed. Therefore, 80 percent of students passed the exam.

14) Choice D is correct

If 17 balls are removed from the bag at random, there will be one ball in the bag. The probability of choosing a brown ball is 1 out of 18. Therefore, the probability of not choosing a brown ball is 17 out of 18 and the probability of having not a brown ball after removing 17 balls is the same.

15) Choice A is correct.

The second digit to the right of the decimal point is in the hundredths place and the third number to the right of the decimal point is in the thousandths place. Since the number in the thousandths place of 0.5749, which is 4, is less than 5, the number 0.5749 should be rounded down to 0.57

16) Choice D is correct

$\dfrac{3}{4} \times 28 = \dfrac{84}{4} = 21$

17) Choice D is correct

Ethan needs an 75% average to pass for five exams. Therefore, the sum of 5 exams must be at lease $5 \times 75 = 375$, The sum of 4 exams is: $68 + 72 + 85 + 90 = 315$.

www.EffortlessMath.com

The minimum score Jason can earn on his fifth and final test to pass is: $375 - 315 = 60$

18) Choice B is correct

6% of the volume of the solution is alcohol. Let x be the volume of the solution.

Then: $6\% \ of \ x = 24 \ ml \Rightarrow 0.06 \ x = 24 \Rightarrow x = 24 \div 0.06 = 400$

19) Choice C is correct

Let x be the smallest number. Then, these are the numbers: $x, x + 1, x + 2, x + 3, x + 4$

$\text{average} = \frac{\text{sum of terms}}{\text{number of terms}} \Rightarrow 36 = \frac{x+(x+1)+(x+2)+(x+3)+(x+4)}{5} \Rightarrow 36 = \frac{5x+10}{5} \Rightarrow$

$180 = 5x + 10 \Rightarrow 170 = 5x \Rightarrow x = 34$

20) Choice C is correct

The fraction $\frac{2}{5}$ can be written as $\frac{2 \times 20}{5 \times 20} = \frac{40}{100}$, which can be interpreted as forty hundredths, or 0.40.

21) Choice A is correct

Write a proportion and solve for the missing number. $\frac{32}{12} = \frac{6}{x} \rightarrow 32x = 6 \times 12 = 72$

$32x = 72 \rightarrow x = \frac{72}{32} = 2.25$

22) Choice D is correct

Dividing 75 by 15%, which is equivalent to 0.15, gives 500.

23) Choice D is correct

Five years ago, Amy was three times as old as Mike. Mike is 10 years now. Therefore, 5 years ago Mike was 5 years. Five years ago, Amy was: $A = 3 \times 5 = 15$, Now Amy is 20 years old: $15 + 5 = 20$

24) Choice C is correct

The average speed of john is: $150 \div 6 = 25$, The average speed of Alice is: $140 \div 4 = 35$

Write the ratio and simplify. $25 : 35 \Rightarrow 5 : 7$

25) Choice A is correct

Let x be the integer. Then: $2x - 12 = 80$, Add 5 both sides: $2x = 92$, Divide both sides by 2: $x = 46$

AFOQT Practice Test 1: Mathematics Knowledge

1) Choice D is correct

Use FOIL method. $(5x + 2y)(2x - y) = 10x^2 - 5xy + 4xy - 2y^2 = 10x^2 - xy - 2y^2$

2) Choice E is correct

To solve absolute values equations, write two equations. $x - 10$ could be positive 4, or negative 4. Therefore, $x - 10 = 4 \Rightarrow x = 14$, $x - 10 = -4 \Rightarrow x = 6$. Find the product of solutions: $6 \times 14 = 84$

3) Choice B is correct

The equation of a line in slope intercept form is: $y = mx + b$. Solve for y.

$4x - 2y = 6 \Rightarrow -2y = 6 - 4x \Rightarrow y = (6 - 4x) \div (-2) \Rightarrow y = 2x - 3$. The slope is 2.

The slope of the line perpendicular to this line is: $m_1 \times m_2 = -1 \Rightarrow 2 \times m_2 = -1 \Rightarrow m_2 = -\frac{1}{2}$.

4) Choice C is correct

Plug in the value of x and y. $x = 3$ and $y = -2$.

$6(x - 2y) + (2 - x)^2 = 6(3 - 2(-2)) + (2 - 3)^2 = 6(3 + 4) + (-1)^2 = 42 + 1 = 43$

5) Choice B is correct

The diagonal of the square is 4. Let x be the side. Use Pythagorean Theorem: $a^2 + b^2 = c^2$

$x^2 + x^2 = 4^2 \Rightarrow 2x^2 = 4^2 \Rightarrow 2x^2 = 16 \Rightarrow x^2 = 8 \Rightarrow x = \sqrt{8}$

The area of the square is: $\sqrt{8} \times \sqrt{8} = 8$

6) Choice B is correct

Isolate and solve for x. $\frac{2}{3}x + \frac{1}{6} = \frac{1}{2} \Rightarrow \frac{2}{3}x = \frac{1}{2} - \frac{1}{6} = \frac{1}{3} \Rightarrow \frac{2}{3}x = \frac{1}{3}$. Multiply both sides by the reciprocal of the coefficient of x. $(\frac{3}{2})\frac{2}{3}x = \frac{1}{3}(\frac{3}{2}) \Rightarrow x = \frac{3}{6} = \frac{1}{2}$

7) Choice B is correct

Use simple interest formula:$I = prt$ ($I =$ interest, $p =$ principal, $r =$ rate, $t =$ time).

$$I = (12,000)(0.045)(2) = 1,080$$

8) Choice D is correct

Simplify. $7x^2y^3(2x^2y)^3 = 7x^2y^3(8x^6y^3) = 56x^8y^6$

9) Choice A is correct

Surface Area of a cylinder $= 2\pi r (r + h)$, The radius of the cylinder is 2 $(4 \div 2)$ inches and its height is 8 inches. Therefore, Surface Area of a cylinder $= 2\pi (2) (2 + 8) = 40 \pi$

10) Choice D is correct

Use the information provided in the question to draw the shape.

Use Pythagorean Theorem: $a^2 + b^2 = c^2$

$50^2 + 120^2 = c^2 \Rightarrow 2,500 + 14,400 = c^2 \Rightarrow 16,900 = c^2 \Rightarrow c = 130$

11) Choice B is correct

Plug in 104 for F and then solve for C.

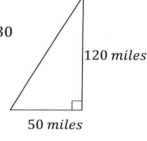

Port A
120 *miles*
50 *miles*

$C = \dfrac{5}{9} (F - 32) \Rightarrow C = \dfrac{5}{9} (104 - 32) \Rightarrow C = \dfrac{5}{9} (72) = 40$

12) Choice A is correct

The width of the rectangle is twice its length. Let x be the length. Then, $width = 2x$

Perimeter of the rectangle is 2 $(width + length) = 2(2x + x) = 72 \Rightarrow 6x = 72 \Rightarrow x = 12$. Length of the rectangle is 12 meters.

13) Choice D is correct

Solve for y. $4x - 2y = 12 \Rightarrow -2y = 12 - 4x \Rightarrow y = 2x - 6$. The slope of the line is 2.

14) Choice C is correct

The formula for the area of the circle is: $A = \pi r^2$,The area is 36π. Therefore:$A = \pi r^2 \Rightarrow 6\pi = \pi r^2$, Divide both sides by π: $36 = r^2 \Rightarrow r = 6$. Diameter of a circle is $2 \times$ radius. Then:

$Diameter = 2 \times 6 = 12$

15) Choice D is correct

$g(x) = -3$, **then** $f\big(g(x)\big) = f(-3) = 2\,(-3)^3 + 5(-3)^2 + 2(-3) = -54 + 45 - 6 = -15$

16) Choice D is correct

Let x be the width of the rectangle. Use Pythagorean Theorem:

$a^2 + b^2 = c^2$

$x^2 + 6^2 = 10^2 \Rightarrow x^2 + 36 = 100 \Rightarrow x^2 = 100 - 36 = 64 \Rightarrow x = 8$

Perimeter of the rectangle $= 2\,(length + width) = 2\,(8 + 6) = 2\,(14) = 28$

17) Choice B is correct

The perimeter of the trapezoid is 40.herefore, the missing side (height) is

$= 40 - 8 - 12 - 6 = 14$. Area of a trapezoid: $A = \frac{1}{2}\,h\,(b_1 + b_2) = \frac{1}{2}\,(14)\,(6 + 8) = 98$

18) Choice D is correct

$f\big(g(x)\big) = 2 \times \left(\frac{1}{x}\right)^3 + 2 = \frac{2}{x^3} + 2$

19) Choice D is correct

Use the information provided in the question to draw the shape.

Use Pythagorean Theorem: $a^2 + b^2 = c^2$

$80^2 + 150^2 = c^2 \Rightarrow 6400 + 22500 = c^2 \Rightarrow 28900 = c^2 \Rightarrow c = 170$

150 miles

Port A

80 miles

20) Choice C is correct

Write the ratio of $5a$ to $2b$. $\frac{5a}{2b} = \frac{1}{10}$. Use cross multiplication and then simplify.

$5a \times 10 = 2b \times 1 \to 50a = 2b \to a = \frac{2b}{50} = \frac{b}{25}$

Now, find the ratio of a to b. $\frac{a}{b} = \frac{\frac{b}{25}}{b} \rightarrow \frac{b}{25} \div b = \frac{b}{25} \times \frac{1}{b} = \frac{b}{25b} = \frac{1}{25}$

21) Choice A is correct

Plug in the value of x in the equation and solve for y. $2y = \frac{2x^2}{3} + 6 \rightarrow 2y =$

$\frac{2(9)^2}{3} + 6 \rightarrow 2y = \frac{2(81)}{3} + 6 \rightarrow 2y = 54 + 6 = 60 \rightarrow 2y = 60 \rightarrow y = 30$

22) Choice D is correct

Since $N = 6$, substitute 6 for N in the equation $\frac{x-3}{5} = N$, which gives $\frac{x-3}{5} = 6$. Multiplying both sides of $\frac{x-3}{5} = 6$ by 5 gives $x - 3 = 30$ and then adding 3 to both sides of $x - 3 = 30$ then,

$x = 33$.

23) Choice C is correct

$b^{\frac{m}{n}} = \sqrt[n]{b^m}$ For any positive integers m and n. Thus, $b^{\frac{3}{5}} = \sqrt[5]{b^3}$

24) Choice B is correct

The total number of pages read by Sara is 3 (hours she spent reading) multiplied by her rate of reading: $\frac{N pages}{hour} \times 3 hours = 3N$

Similarly, the total number of pages read by Mary is 4 (hours she spent reading) multiplied by her rate of reading: $\frac{M pages}{hour} \times 4 hours = 4M$ the total number of pages read by Sara and Mary is the sum of the total number of pages read by Sara and the total number of pages read by Mary: $3N + 4M$.

25) Choice E is correct

The initial deposit earns 2 percent interest compounded annually. Thus, at the end of year 1, the new value of the account is the initial deposit of \$150 plus 2 percent of the initial deposit: $\$150 + \frac{2}{100}(\$150) = \$150(1.02)$.

Since the interest is compounded annually, the value at the end of each succeeding year is the sum of the previous year's value plus 2 percent of the previous year's value. This is equivalent to multiplying the previous year's value by 1.02. Thus, after 2 years, the value will be $\$150(1.02)(1.02) = \$(150)(1.02)^2$; and after 3 years, the value will be $(150)(1.02)^3$; and after n years, the value will be $(150)(1.02)^n$. Therefore, in the formula for the value for Sara's account after n years $(100)(x)^n$, the value of x is 1.02.

AFOQT Practice Test 2: Arithmetic Reasoning

1) Choice E is correct
2 weeks = 14 days, Then: $14 \times 6 = 84$ hours, $84 \times 60 = 5,040$ minutes

2) Choice E is correct
$distance = speed \times time \Rightarrow$ time $= \frac{distance}{speed} = \frac{340+340}{50} = 13.6$

(Round trip means that the distance is 680 miles)

The round trip takes 13.6 hours. Change hours to minutes, then: $13.6 \times 60 = 816$

3) Choice C is correct
$60 - 42 = 18$ male students. $\frac{18}{60} = 0.3$, Change 0.3 to percent $\Rightarrow 0.3 \times 100 = 30\%$

4) Choice C is correct
$averag = \frac{sum}{total}$, Sum $= 7 + 9 + 22 + 28 + 28 + 30 = 124$

Total number of numbers = 9 $\qquad \frac{124}{6} = 20.67$

5) Choice A is correct
Emma's three best times are 54, 57, and 57. The average of these numbers is:
$average = \frac{sum}{total}$

Sum $= 54 + 57 + 57 = 168$. Total number of numbers = 3 $\quad average = \frac{168}{3} = 56$

6) Choice A is correct
The area of a 15 feet x 15 feet room is 225 square feet. $15 \times 15 = 225$
7) Choice B is correct
$1.303572 \times 1000 = 1303.572$

8) Choice D is correct
The factors of 50 are: {1, 2, 5, 10, 25, 50}. 15 is not a factor of 50.

9) Choice A is correct
4 percent of 25 is: $25 \times \frac{4}{100} = 1$, Emma's new rate is 26. $25 + 1 = 26$

10) Choice A is correct
Emily = Lucas, Emily = 4 Mia ⇒ Lucas = 4 Mia, Lucas = Mia + 21, then:
Lucas = Mia + 21 ⇒ 4 Mia = Mia + 21. Remove 1 Mia from both sides of the equation. Then: 3 Mia = 21 ⇒ Mia = 7

11) Choice C is correct
12 days, 12 × 5 = 60 hours, 60 × 60 = 3,600 minutes

12) Choice A is correct
Sum = 22 + 34 + 16 + 20 = 92, $average = \frac{92}{4} = 23$

13) Choice B is correct
Perimeter of a rectangle = 2 × length + 2 × width = 2 × 90 + 2 × 30 = 180 + 60 = 240

14) Choice B is correct
$Speed = \frac{distance}{time}$, $16.2 = \frac{distance}{2.1} \Rightarrow distance = 16.2 \times 2.1 = 34.02$

Rounded to a whole number, the answer is 34.

15) Choice B is correct
Let's review the choices provided and find their sum.

A. 20 × 6 = 120
B. 26 × 6 = 144 ⇒ is greater than 120 and less than 180
C. 30 × 6 = 180
D. 34 × 6 = 204
E. 34 × 6 = 228

Only choice B gives a number that is greater than 120 and less than 180.

16) Choice C is correct
$\frac{1\ hour}{15\ coffees} = \frac{x}{1500} \Rightarrow 15 \times x = 1 \times 1,500 \Rightarrow 15x = 1,500$, $x = 100$

It takes 100 hours until she's made 1,500 coffees.

17) Choice A is correct
120 − 12 = 108, $\frac{108}{12} = 9$

18) Choice C is correct
$percent\ of\ change = \frac{change}{original\ number}$, 7.75 − 7.50 = 0.25

$percent\ of\ change = \dfrac{0.25}{7.50} = 0.0333$ $\Rightarrow 0.0333 \times 100 = 3.33\%$

19) Choice A is correct

Write a proportion and solve. $\dfrac{\frac{1}{2}inches}{4.5} = \dfrac{1\ mile}{x}$

Use cross multiplication, then: $\dfrac{1}{2}x = 4.5 \rightarrow x = 9$

20) Choice B is correct

Two candy bars costs 50¢ and a package of peanuts cost 75¢ and a can of cola costs 50¢. The total cost is: $50 + 75 + 50 = 175$, 175 is equal to 7 quarters. $7 \times 25 = 175$

21) Choice C is correct

Every day the hour hand of a watch makes 2 complete rotation. Thus, it makes 16 complete rotations in 8 days. $2 \times 8 = 16$

22) Choice E is correct

$\sqrt{81} \times \sqrt{25} = 9 \times 5 = 45$

23) Choice A is correct

$2y + 4y + 2y = -24$ $\Rightarrow 8y = -24 \Rightarrow y = -\dfrac{24}{8} \Rightarrow y = -3$

24) Choice D is correct

$2\dfrac{2}{3} - 1\dfrac{5}{6} = 2\dfrac{4}{6} - 1\dfrac{5}{6} = \dfrac{16}{6} - \dfrac{11}{6} = \dfrac{5}{6}$

25) Choice C is correct

To convert a decimal to percent, multiply it by 100 and then add percent sign (%).

$0.023 \times 100 = 2.30\%$

AFOQT Practice Test 2: Mathematics Knowledge

1) Choice D is correct
Use FOIL (First, Out, In, Last) method.

$(x + 7)(x + 5) = x^2 + 5x + 7x + 35 = x^2 + 12x + 3$

2) Choice D is correct
In scientific notation form, numbers are written with one whole number times 10 to the power of a whole number. Number 670,000 has 6 digits. Write the number and after the first digit put the decimal point. Then, multiply the number by 10 to the power of 5 (number of remaining digits). Then: $670,000 = 6.7 \times 10^5$

3) Choice C is correct
Perimeter of a triangle = side 1 + side 2 + side 3 = 25 + 25 + 25 = 75

4) Choice A is correct
From the choices provided, 36, 48 and 54 are divisible by 6. From these numbers, 54 is the biggest.

5) Choice A is correct
Oven 1 = 4 oven 2. If Oven 2 burns 3 then oven 1 burns 12 pizzas. 3 + 12 = 15

6) Choice C is correct
An obtuse angle is an angle of greater than 90° and less than 180°.

7) Choice E is correct
The description $8 + 2x$ *is* 16 more than 20 can be written as the equation $8 + 2x = 16 + 20$, which is equivalent to $8 + 2x = 36$. Subtracting 8 from each side of $8 + 2x = 36$ gives $2x = 28$. Since $6x$ is 3 times $2x$, multiplying both sides of $2x = 28$ by 3 gives $6x = 84$.

8) Choice E is correct
$\frac{3}{5} \times 30 = \frac{90}{5} = 18$

9) Choice D is correct
The slop of line A is: $m = \frac{y_2 - y_1}{x_2 - x_1} = \frac{3-2}{4-3} = 1$, Parallel lines have the same slope and only choice D ($y = x$) has slope of 1.

10) Choice A is correct

Diameter = 16, then: Radius = 8, Area of a circle = $\pi r^2 \Rightarrow A = 3.14(8)^2 = 200.96$

11) Choice D is correct

$y = (-3x^3)^2 = (-3)^2(x^3)^2 = 9x^6$

12) Choice C is correct

Let's review the choices provided. Put the values of x and y in the equation.

A. $(1, 2) \Rightarrow x = 1 \Rightarrow y = 2$ This is true!

B. $(-2, -13) \Rightarrow x = -2 \Rightarrow y = -13$ This is true!

C. $(3, 18)$ $\Rightarrow x = 3 \Rightarrow y = 12$ This is not true!

D. $(2, 7) \Rightarrow x = 2 \Rightarrow y = 7$ This is true!

E. $(5, 3) \Rightarrow x = 5 \Rightarrow y = 22$ This is not true!

13) Choice D is correct

In the question, there are two equations and three variables. Therefore, it cannot be determined from the information given.

14) Choice C is correct

Use distance formula:

$$d = \sqrt{(x_1 - x_2)^2 + (y_1 - y_2)^2} = \sqrt{(1 - (-2))^2 + (3 - 7)^2} \qquad \sqrt{9 + 16} = \sqrt{25} = 5$$

15) Choice B is correct

$x^2 - 81 = 0 \quad \Rightarrow \quad x^2 = 81 \quad \Rightarrow x$ could be 9 or –9.

16) Choice C is correct

Area of a rectangle = width × length = $160 \times 200 = 3{,}200$

17) Choice D is correct

The angle x and 60 are complementary angles. Therefore: $x + 60 = 180$, $180° - 60° = 120°$

18) Choice C is correct

Two triangles ΔBAE and ΔBCD are similar. Then:

$\frac{AE}{CD} = \frac{AB}{BC} = \frac{6}{9}$ and $AC = 24$. Let's put x for AB. Then: $AC = AB + BC = x + (24 - x)$.

$\frac{6}{9} = \frac{x}{24 - x} \rightarrow 144 - 6x = 9x \rightarrow 15x = 144 \rightarrow x = 9.6$

19) Choice B is correct

To solve this problem, first recall the equation of a line: $y = mx + b$

Where, $m = slope$ and $y = y - intercept$, remember that slope is the rate of change that occurs in a function and that the $y -$intercept is the y value corresponding to $x = 0$.

Since the height of John's plant is 6 inches tall when he gets it. Time (or x) is zero. The plant grows 4 inches per year. Therefore, the rate of change of the plant's height is 4. The $y -$intercept represents the starting height of the plant which is 6 inches.

20) Choice B is correct

The cube of $4 = 4 \times 4 \times 4 = 64$, $\frac{1}{4} \times 64 = 16$

21) Choice C is correct

From the list of numbers, 11, 13, and 19 are prime numbers. Their sum is:

$11 + 13 + 19 = 43$

22) Choice B is correct

$x^2 = 144 \rightarrow x = 12$ (positive value)   Or $x = -12$ (negative value)

Since x is positive, then: $f(144) = f(12^2) = 3(12) + 5 = 36 + 5 = 41$

23) Choice A is correct

Two Angles are supplementary when they add up to 180 degrees.

$135° + 45° = 180°$

24) Choice A is correct

$3.5x - 17.5 < 2x - 5 - 3.5x \rightarrow$ Combine like terms: $3.5x - 17.5 < -1.5x - 5 \rightarrow$

Add $1.5x$ to both sides: $5x - 17.5 < -5 \rightarrow$ Add 17.5 to both sides of the inequality.

$5x < 12.5 \rightarrow$ Divide both sides by 5. $\rightarrow \frac{5}{5}x < \frac{12.5}{5} \rightarrow x < 2.5$

25) Choice C is correct

$5(2x^6)^3 \Rightarrow 5 \times 2^3 \times x^{18} = 40x^{18}$

"Effortless Math Education" Publications

Effortless Math authors' team strives to prepare and publish the best quality AFOQT Mathematics learning resources to make learning Math easier for all. We hope that our publications help you learn Math in an effective way and prepare for the AFOQT test.

We all in Effortless Math wish you good luck and successful studies!

Effortless Math Authors

www.EffortlessMath.com

… So Much More Online!

- ❖ FREE Math lessons
- ❖ More Math learning books!
- ❖ Mathematics Worksheets
- ❖ Online Math Tutors

Need a PDF version of this book?

Visit www.EffortlessMath.com

Receive the PDF version of this book or get another FREE book!

Thank you for using our Book!

Do you LOVE this book?

Then, you can get the PDF version of this book or another book absolutely FREE!

Please email us at:

info@EffortlessMath.com

for details.

Made in the USA
Las Vegas, NV
06 December 2020